Kuhn's
Structure of Scientific Revolutions
at Fifty

Kuhn's
Structure of Scientific Revolutions
at Fifty

Reflections on a Science Classic

Edited by
ROBERT J. RICHARDS and
LORRAINE DASTON

University of Chicago Press
Chicago and London

Robert J. Richards is the Morris Fishbein Distinguished Service
Professor of the History of Science and Medicine; professor in
the Departments of History, Philosophy, and Psychology and in
the Committee on Conceptual and Historical Studies of Science;
and director of the Fishbein Center for the History of Science and
Medicine, all at the University of Chicago.

Lorraine Daston is director of the Max Planck Institute for
the History of Science in Berlin, and visiting professor in the
Committee on Social Thought at the University of Chicago.

The University of Chicago Press, Chicago 60637
The University of Chicago Press, Ltd., London
© 2016 by The University of Chicago
All rights reserved. Published 2016.
Printed in the United States of America

24 23 22 21 20 19 18 17 16   1 2 3 4 5

ISBN-13: 978-0-226-31703-8 (cloth)
ISBN-13: 978-0-226-31720-5 (paper)
ISBN-13: 978-0-226-31717-5 (e-book)

DOI: 10.7208/chicago/9780226317175.001.0001

Library of Congress Cataloging-in-Publication Data

Kuhn's Structure of Scientific Revolutions at fifty : reflections on a
science classic / edited by Robert J. Richards and Lorraine Daston.
        pages ; cm
    Includes bibliographical references and index.
    ISBN 978-0-226-31703-8 (cloth : alk. paper) — ISBN 978-0-226-
31720-5 (pbk. : alk. paper) — ISBN 978-0-226-31717-5 (ebook)
1. Kuhn, Thomas S. Structure of scientific revolutions.   2. Science—
Philosophy.   3. Science—History.   I. Richards, Robert J. (Robert
John), 1942– editor.   II. Daston, Lorraine, 1951– editor.
    Q175.K84 2016
    501—dc23

                                              2015017230

♾ This paper meets the requirements of ANSI/NISO Z39.48-1992
(Permanence of Paper).

CONTENTS

Portrait of Thomas Kuhn in his late teens

INTRODUCTION

You must have a special circle of Hell reserved for authors who write to request special treatment. Let me try to persuade you to be slow in assigning
me to it. I would not write this sort of letter if I did not think the view of science developed in my manuscript might prove particularly important.[1]

Like so many authors, Thomas Kuhn, then associate professor of the history of science at the University of California, Berkeley, had come to regret
an earlier publishing commitment. When he had agreed some eight years
previous in 1953 to contribute an article to the *International Encyclopedia of
Unified Science*[2] titled "The Structure of Scientific Revolutions," Kuhn was a
freshly minted assistant professor at Harvard in General Education and the
history of science, subjects in which he held no degrees.[3] He was flattered,
no doubt, by the invitation to contribute to a venture so closely associated
with the Vienna Circle in exile.[4] His article would appear in the best of
philosophical company: Ernest Nagel on the principles of probability theory, John Dewey on the theory of valuation, Rudolf Carnap of the foundations of logic and mathematics. But as the years (and the original deadline)
went by, the manuscript waxed, and so did the author's ambitions. His first
book, *The Copernican Revolution: Planetary Astronomy in the Development of
Western Thought*,[5] had enjoyed sufficient success to net him a joint appointment in the Departments of History and Philosophy at Berkeley; the re-
nown of the *Encyclopedia* had in the meantime waned (Kuhn claimed that
his primary audience of "historians, sociologists, and scientists" had never
even heard of it). The "special treatment" that Kuhn was requesting in his
1961 letter to Carroll Bowen of the University of Chicago Press was not
exactly a release from what he considered to be "at the very least, a strong

moral commitment" to the *Encyclopedia* but rather "simultaneous publication in another format."[6]

This is the book that became *The Structure of Scientific Revolutions*,[7] a book judged by the *Times Literary Supplement* to be among the twenty most influential books published in the second half of the twentieth century.[8] The press had obliged, and with astonishing and gratifying alacrity: as Kuhn recalled a couple of years later, "Three days after I mailed a draft of my manuscript with a covering letter asking for a special publication arrangement, I had a telephone call from Chicago telling me that the Press was immensely excited to have the manuscript and that the special conditions I asked would, though unprecedented, present no problem at all."[9] Perhaps the excitement was not quite so immense as the author fondly imagined: the initial print run was prudently set at 1,000 clothbound and 2,000 paperback copies, priced at $4.00 and $3.00, respectively.[10] Even the author confessed on the press's author's questionnaire (which has remained remarkably unchanged for over a half-century) that he had "no good ideas" in response to the standard question about which professors and institutions might be interested in using the book for classroom instruction.[11] At first, these modest estimates were confirmed by sales: a respectable but not spectacular 919 copies sold in the first year and another 774 copies a year later. But by the early 1970s, the book was selling over 40,000 copies a year and was being assigned in courses across university campuses all over the country and beyond. The first edition sold a total of 88,538 paperback copies and the second edition (up to mid-1987), 546,455.[12] The book was a sensation, its leading ideas hotly debated among natural and social scientists as well as humanists and parodied in *New Yorker* cartoons. Even its author, buoyantly self-confident about the significance of his book ("the book is more ambitious and important than most of those you print," he assured the marketing department of the press in a pitch for more than the usual publicity[13]), must have been taken by surprise (and sometimes aback) by his celebrity.

Biographical Background

Not that Kuhn was a man of small academic ambition.[14] Born July 18, 1922, in Cincinnati, Ohio, Thomas Samuel Kuhn grew up in Manhattan and Croton-on-Hudson, attending the Taft School in Connecticut and later Harvard College, whence he graduated summa cum laude with a concentration in physics in 1943, after only three years of study while editing the student newspaper the *Harvard Crimson* on the side. Immediately thereafter, he plunged into war-related work on radar, which took him to France and then Germany at the end of the war. He returned to Harvard as a graduate student in physics in the fall semester of 1945, with John Hasbrouck Van Vleck as his thesis advisor. While there, he secured special permission to take philosophy (not history) courses, having been much impressed by an undergraduate encounter with the writings of Kant. Although he clearly was entertaining serious interests outside of physics, he completed his dissertation with dispatch and confessed later to dreaming of a Nobel Prize someday (surely his name was and is legion among physics graduate students).

But the course of his academic pursuits had begun to shift decisively, largely through the intervention of the then-president of Harvard, chemist James Bryant Conant. It was Conant who saw to it that Kuhn was awarded a junior fellowship in the Harvard Society of Fellows in 1948, and it was Conant who recruited Kuhn to teach in a new General Education course that taught non–scientifically inclined Harvard undergraduates about how scientists think through a series of historical case studies. Conant had been a prominent scientific administrator during World War II and believed that a scientifically literate citizenry was essential to the future of American democracy. Kuhn was assigned the task of preparing the case study in the history of mechanics from Aristotle to Galileo, a task that soon had him reading the historical work of the German medievalist Anneliese Maier and the Russian-French philosopher and historian Alexandre Koyré, figures to whom, along with French historian of chemistry Hélène Metzger, philosopher Émile Meyerson, and American historian of ideas A. O. Lovejoy, he confessed his intellectual debt in the preface of *The Structure of Scientific Revolutions*.

In his own telling, however, it was reading Aristotle in the summer of 1947 that sparked Kuhn's historical epiphany. With the sudden all-at-onceness that Kuhn later likened to a Gestalt switch, he realized that Aristotle's account of motion was not simply wrong; it was about something else entirely. Aristotle, Kuhn realized, understood motion not just as displacement from point A to point B but as a special case of all kinds of change, from the maturing of organisms to the succession of the seasons. This habit of reading the past sympathetically, in its own terms and context, made Kuhn permanently wary of what he called (borrowing the phrase from the British political historian Herbert Butterfield) "Whig" history of science: narrating the history of science teleologically, as a triumphal conquest of error culminating in present scientific beliefs.[15] Thereafter, Kuhn struggled to find other narratives that would do justice to scientific advances without succumbing to the siren song of the conventional (among scientists as well as historians and philosophers) story of cumulative progress approaching ever nearer to eternal truths. He flirted with the analogy to Darwinian evolution, although it accorded ill with his scheme of periodic revolutions.

Although the germs of the ideas for *Structure* were planted in the context of Conant's General Education course at Harvard (Kuhn's *The Copernican Revolution* also began as one of the case studies), the manuscript took shape in California, where Kuhn quickly rose through the academic ranks at Berkeley after his joint appointment as assistant professor in the History and Philosophy Departments in 1956 and where he spent the academic year 1958–59 as a fellow at the Stanford Center for Advanced Study in the Behavioral Sciences. In the early 1960s, Kuhn undertook an oral history of quantum mechanics in collaboration with John Heilbron, Paul Foreman, and Lini Allen, resulting in *Sources for the History of Quantum Physics* (1967). Once *Structure* came out in 1962, Kuhn turned his attention to the history of quantum mechanics, a project that sent him to Copenhagen in 1962–63 (where he interviewed Niels Bohr), and eventually resulted in the book Kuhn himself considered his finest work, *Black-Body Radiation and the Quantum Discontinuity, 1894–1912* (1978). By the time it was published, Kuhn was about to leave Princeton, where he had accepted an appointment in 1964, for MIT, where he spent the remainder of his academic career, from 1979 until his death in 1996. *Structure* had brought him international fame and innumerable honors, but he apparently considered it an essay, a preliminary attempt in need of expansion and improvement. His final, unfinished book manuscript was a philosophical and linguistic exploration of problems raised by the controversy over *Structure*, especially among philosophers provoked by Kuhn's insistence that successive scientific paradigms were in some sense "incommensurable."[16] Despite the fact that all of his training and most of his publications concerned the physical sciences, he once again drew inspiration from biology, this time from the idea of natural kinds and taxonomic trees.

The Reception of Kuhn's *Structure*

Kuhn's *Structure of Scientific Revolutions* had a gentle birth but a traumatic childhood. When Rudolf Carnap, an editor of the *International Encyclopedia of Unified Science* and philosopher at the University of Chicago, received Kuhn's completed manuscript, he was undoubtedly relieved, since two previous historians who had been solicited to contribute a monograph on the history of science had failed to deliver. Carnap congratulated Kuhn on the aptness of his essay and especially singled out a conception that he thought conformed to his own developing ideas about scientific language, namely, Kuhn's suggestion that theory change paralleled species change as explained in Darwin's evolutionary theory.[17] Theory change, in Carnap's assessment, was the improvement of an instrument not a better approximation to an ultimately true theory. Kuhn had deployed the evolutionary comparison only briefly in the last part of the essay, but without any proposal that paradigms gradually replaced one another in an evolutionary fashion, as Darwin's theory might have suggested.

The University of Chicago Press, under Carnap's guidance, chose Norwood Russell Hanson as one of the referees for Kuhn's manuscript.[18] The choice of Hanson, a prominent philosopher of science at Indiana University, was not accidental. Like Kuhn, Hanson had deployed examples from Gestalt psychology to exemplify comparable ideas about the theory-ladenness of observation; he even used the term "paradigm" in rather similar ways, if without systematic intent.[19] Hanson judged the monograph "a work of surpassing merit, scholarship, and creative ingenuity," but he did have a salient criticism: in Popperian fashion he wanted to know what would falsify the thesis that scientific revolutions resulted from a paradigm shift. Since Kuhn did not provide independent criteria for a revolution or a shift, the thesis appeared to be analytic and not an empirical assertion: a revolution occurred when a paradigm shift took place, and a shift took place when a revolution occurred.[20] Popperian considerations would continue to trouble Kuhn's theory.

In the first three years after publication, the book garnered a generous swathe of reviews for such a slim production with a drab, gray cover—reviews from philosophers of science, historians of science, and scientists. Immediately on its publication, the historian Charles Coulston Gillispie of Princeton wrote a long, leading review in *Science* of "this very bold venture." Like Carnap, he emphasized what he took to be Kuhn's "Darwinian historiography" as congenial to his own.[21] In the *American Historical Review*, Maria Boas Hall gave the book a gentle recommendation, finding the influence of George Sarton flavoring Kuhn's analysis—though one suspects Sarton would have found it bitter tasting.[22] The sociologist Bernard Barber characterized the book as "an essay in the sociology of scientific discovery," though he wished Kuhn had dealt more with the "external factors" influencing scientific change.[23] The physicist David Bohm regarded the book as "without doubt one of the most interesting and significant contributions that has been made in recent years in the field of the history and philosophy of science."[24] His own heterodox interpretation of quantum theory undoubtedly led him to appreciate Kuhn's discontinuity thesis.

While these early reviewers tended to greet Kuhn as a fellow traveler, several philosophers of science were less tolerant of the abandonment of a more traditional picture of scientific progress; they exposed contradictions, inconsistencies, and loose ends in Kuhn's conception. Dudley Shapere at the University of Chicago provided a token last cigarette—allowing that it was an "important" book—before he fired relentlessly into the heart of Kuhn's thesis, aiming to show that the term "paradigm" covered a range of factors yet without a clearly specifiable meaning—a charge that would later be made even more deadly. Shapere delighted in exposing the logical conundrums endemic to the concept of incommensurability, scattering Kuhn's conflicting statements about progress through the last pages of a brutal review.[25] Mary Hesse, at the University of Cambridge, also found the book "important," though she trod more gently on the several major inconsistencies she found in the argument.[26] H. V. Stopes-Roe, a little-published philosopher at Birmingham University, did not even grant a polite condescension before enumerating the mind-stopping contradictions he found in Kuhn's thesis.[27]

The reactions of reviewers were prelude to an event that set the attitude for a significant portion of Kuhn's readers. In 1965, a meeting of the International Colloquium in the Philosophy of Science was held in London titled "Criticism and the Growth of Knowledge," a theme central to the convictions of Karl Popper, who presided as the meeting's chair. Kuhn was invited to be a principal speaker, along with Paul Feyerabend (Kuhn's colleague at Berkeley) and Imré Lakatos, both of growing reputation for taking a deviant stance. Neither of the latter two finished their papers on time for the conference, so Kuhn's contribution became a central focus of the discussants, who included the aforementioned along with Stephen Toulmin, L. Pearce Williams, Margaret Masterman, and John Watkins. All of the papers were published four years later, after the tardy authors finished their contributions (Lakatos's "Falsification and the Methodology of Scientific Research Programmes" ran to over one hundred pages, and became itself quite influential and more dependent on Kuhn than the author admitted). Kuhn was allowed to answer his critics in the published volume *Criticism and the Growth of Knowledge.*[28]

Kuhn came ready for combat. After pointing out several features of his theory consonant with those of Popper, he set his sights on the principle of falsification itself, arguing that only an observation *statement*, rather than a physical observation of the kind Popper thought crucial, could refute a general theory. After all, we might decide that an actual observation was mismeasured, that it was actually irrelevant, that something interfered with the outcome, and so on.[29] Popper simply ignored the objection, but did grant that Kuhn made an important distinction between normal science and extraordinary, paradigm-shifting science. Popper, however, knew how to avoid a critical objection: he contended that so-called normal science arose from poor training, in which tyros had not been taught to exercise their critical faculties; in proper science, there were only degrees of difference between smaller changes in theory and groundbreaking changes. What really disturbed Popper—something that became a burr under the saddle of many high-riding philosophers and a stimulus for the new sociologists of science—was Kuhn's explicit contention that the logic of falsificationism cannot explain paradigm change but that the explanation must, "in the final analysis, be psychological or sociological."[30] Popper simply rejected the appeal to social factors to explain fundamental change as the merest "relativism."[31] The charge of relativism would reverberate through the philosophical literature—given force by Lakatos's epitome of Kuhn's work as "science by mob-rule"—and it continued to bedevil Kuhn to the end. Masterman's analysis struck a comparably tender spot. She found twenty-one different meanings for "paradigm" in the book, which is to say, there was no clear meaning at all.[32] Kuhn had to grant that his treatment was "badly confused." He also recognized that he had to salvage the idea of scientific progress, a concept he did not really wish to abandon. As a solution to the problem of progress, he came to rely more and more on the conception he had mentioned only in passing in his book but that others had emphasized: "my view of scientific development," he declared, "is fundamentally evolutionary."[33] The definition of paradigm and the problem of progress would occupy the epilogue of the second edition of *Structure*.

While philosophers through the 1970s continued to confront Kuhn forcefully on the issue of incommensurability, the new movement of social constructionism looked to Kuhn as having established the role of social factors in major scientific change. Barry Barnes, in a little book that echoed the London conference volume—*Interests and the Growth of Knowledge*—maintained that Kuhn had "demonstrate[d] that fundamental theoretical transitions in science are not simply rational responses to increased knowledge of reality."[34] Sociology of science need not remain at the sidelines of knowledge formation, rather under Kuhn it could become a major player in the explanation of scientific development.

A survey of some 1,785 journals in history, history of science, philosophy, social sciences, and biology shows that articles mentioning Kuhn's

Structure climbed steadily from 1962 to about 1980 and then leveled off till about 2000, with a gentle decline thereafter—with the exception of mention in journals of biology, which, after reaching a plateau, maintained a constant number of citations through 2010.[35]

The Occasion of this Volume

These days it is barely imaginable that a manuscript of approximately fifty thousand words with no footnotes or bibliography[36] by a relatively unknown scholar would ever make it past the straight and narrow gate of a reputable academic press, much less that the publishers would so graciously accede to the demanding author's every wish (including a belated request that additional copies be sent to reviewers in the fields of psychology, psychoanalysis, and social science, even though Kuhn ingenuously admitted that he had "no notion of what the corresponding journals are"[37]) and whim (concerning the customary index, Kuhn queried hopefully, "Does a volume of this sort need an index? I wouldn't be surprised but I'm damned if I know how one would devise one? Paradigm, 11–74, *passim*"[38]). Kuhn had good reason to be pleased with his publishers.

The University of Chicago Press was first the indulgent (or shrewd, depending on one's point of view) and later (as sales ballooned) the grateful publisher of *The Structure of Scientific Revolutions*, and so it seemed entirely appropriate that it sponsor an event to commemorate the fiftieth anniversary of its publication in Chicago in December 2012, in collaboration with the Fishbein Center for the History of Science and Medicine at the University of Chicago and the Max Planck Institute for the History of Science in Berlin. The editors are especially grateful to Karen Darling and Christie Henry of the press and to Special Collections at the Regenstein Library of the University of Chicago for putting together an exhibition of correspondence relating to the book's publication and multiple editions and translations to accompany the conference, as well as for permission to quote materials from the archives of the press.

Speakers at the conference were given complete freedom as to what aspect of Kuhn's book they wished to address—its prehistory, its claims, its influence, its current significance—so the resulting papers are probably a fair sample of the diverse concerns that the book still raises for historians, philosophers, and sociologists. Some of the papers situate the book in the context of Kuhn's own intellectual biography: his experiences as a physics graduate student doing quintessential "normal science" (Peter Galison), his close relationships with psychologists both before and after the publication of *Structure* (David Kaiser), the Cold War framework of concepts such as "worldview" and "paradigm" (George Reisch). Andrew Abbott makes the citation history of the book the occasion for an analysis of its shifting audiences—and changes in scholarly reading habits. Other papers address the import of key Kuhnian concepts both then and now: "revolution" (Daniel Garber), "scientific community" (M. Norton Wise), and, inevitably, the elusive yet still enticing "paradigm" (Angela Creager, Lorraine Daston, and Ian Hacking).

All of the participants in the two-day conference were struck and perhaps somewhat startled by the intensity of the discussion. At fifty, *The Structure of Scientific Revolutions* still engages, excites, and occasionally enrages its readers. This is grounds for surprise. At least among historians, philosophers, and sociologists of science, words like "paradigm" and "incommensurability" nowadays have a quaint ring to them. Shouting matches over the alleged irrationality of the resolution of scientific revolutions no longer enliven scholarly meetings. We are all post-Kuhnian—yet no one is indifferent to Kuhn's provocative essay. This volume is a testament to the enduring power of the book to stimulate long after it has ceased to be the center of raging disputes—and also to the courage and canniness of a press that knew a good thing when it saw one.

Notes

1. Thomas S. Kuhn to Carroll G. Bowen, Berkeley, California, June 18, 1961, University of Chicago Press Records, Box 278, Folder 4, Special Collections, University of Chicago Library.

2. On the history of the Encyclopedia, see George A. Reisch, "Planning Science: Otto Neurath and the International Encyclopedia of Unified Science," *British Journal for the History of Science* 27 (1994): 153–75; Reisch, "A History of the International Encyclopedia of Unified Science" (Ph.D. diss., University of Chicago, 1995).

3. Karl Hufbauer, "From Student of Physics to Historian of Science: T. S. Kuhn's Education and Early Career, 1940–1958," *Physics in Perspective* 14 (2012): 421–78.

4. The invitation seems to have been secured on Kuhn's behalf by the physicist and philosopher Philipp Frank, an Austrian émigré with close ties to the Vienna Circle: Joel Isaac, *Working Knowledge: Making the Human Sciences from Parsons to Kuhn* (Cambridge, MA: Harvard University Press, 2012), 220.

5. Thomas S. Kuhn, *The Copernican Revolution: Planetary Astronomy in the Development of Western Thought* (Harvard University Press, 1957).

6. Thomas S. Kuhn to Carroll G. Bowen, Berkeley, California, June 18, 1961.

7. Thomas S. Kuhn, *The Structure of Scientific Revolutions* (University of Chicago Press, 1962). The first two editions prominently featured the book's association with the *International Encyclopedia of Unified Science* in the front material; these acknowledgements of the book's prehistory disappeared from the layout of subsequent editions.

8. Paul Hoyningen-Huene, "Thomas S. Kuhn," *Journal for General Philosophy. Zeitschrift für allgemeine Wissenschaftstheorie* 28(1997): 235–56, on 242.

9. Thomas S. Kuhn to Sandra Carroll, Copenhagen, March 13, 1963, University of Chicago Press Records, Box 278, Folder 4, Special Collections, University of Chicago Library.

10. Book Estimate and Release, *The Structure of Scientific Revolutions*, Contract Nr. 6083, [3?]–25–62, University of Chicago Press Records, Box 278, Folder 4, Special Collections, University of Chicago Library

11. In reply to the question about which college courses might adopt the book as a text, Kuhn explained rather lamely that "it's not a text for any course but well might be assigned, if cheap enough, as collateral reading in phil. of science, etc." (Author Questionnaire, Thomas S. Kuhn, *The Structure of Scientific Revolutions*, March 25, 1962, University of Chicago Press Records, Box 278, Folder 4, Special Collections, University of Chicago Library).

12. In the same period, 3,938 clothbound copies of the first edition and 8,819 of the second edition were sold (Publication information on first and second editions of Thomas S. Kuhn, *The Structure of Scientific Revolutions*, University of Chicago Press Records, Box 278, Folder 4, Special Collections, University of Chicago Library.

13. Thomas S. Kuhn to Sandra Carroll, Copenhagen, March 13, 1963.

14. The following condensed biographical account draws on the following sources, which provide considerably more detail concerning Kuhn's personal and professional development: T. S. Kuhn, "A Discussion with Thomas S. Kuhn," *Neusis: Journal for the History and Philosophy of Science and Technology*, 6 (1997): 143–98; Paul Hoyningen-Huene, *Reconstructing Scientific Revolutions: Thomas S. Kuhn's Philosophy of Science* (Chicago: University of Chicago Press, 1993); J. L. Heilbron, "Thomas Samuel Kuhn, 18 July 1922–17 June 1996," *Isis* 89 (1998): 505–15; Jed Z. Buchwald and George E. Smith, "Thomas S. Kuhn, 1922–1996," *Philosophy of Science* 64 (1997): 361–76; Hoyningen-Huene, "Thomas S. Kuhn."

15. Thomas S. Kuhn, "History of Science," in *International Encyclopedia of the Social Sciences*, ed. D. L. Sills (London: Macmillan and Free Press, 1972), 13: 74–83.

16. See the interview with Kuhn in Giovanna Borradori, *The American Philosopher: Conversations with Quine, Davidson, Putnam, Nozick, Danto, Rorty, Cavell, MacIntyre, and Kuhn* (Chicago: University of Chicago Press, 1994), 153–67, esp. 161–66.

17. Rudolf Carnap to Thomas Kuhn (April 28, 1962), transcribed by George Reisch in his "Did Kuhn Kill Logical Empiricism?" *Philosophy of Science* 58 (1991): 264–77.

18. The other reviewer, Roger Hancock (a philosopher at Chicago), gave it an approving but a brief, perfunctory evaluation.

19. See, for example, Norwood Russell Hanson, *Patterns of Discovery* (Cambridge: Cambridge University Press, [1958] 1972), 1, 11–14, 91.

20. Norwood Russell Hanson, "Review of Kuhn Manuscript," University of Chicago Press Records, Box 278, Folder 4, Special Collections, University of Chicago Library.

21. Charles C. Gillispie, "Review of *The Structure of Scientific Revolutions*," *Science* 138 (1962): 1251–53.

22. Maria Boas Hall, "Review of The Structure of Scientific Revolutions," *American Historical Review* 68 (1963): 700–701.

23. Bernard Barber, "Review of The Structure of Scientific Revolutions," *American Sociological Review* 28 (1963): 298–99.

24. David Bohm, "Review of The Structure of Scientific Revolutions," *Philosophical Quarterly* 14 (1964): 377–79.

25. Dudley Shapere, "Review of The Structure of Scientific Revolutions," *Philosophical Review* 70 (1964): 383–94.

26. Mary Hesse, "Review of *The Structure of Scientific Revolutions*," *Isis* 54 (1963): 286–87.

27. H. V. Stopes-Roe, "Review of The Structure of Scientific Revolutions," *British Journal for the Philosophy of Science* 15 (1964): 158–61.

28. Imre Lakatos and Alan Musgrave, eds., *Criticism and the Growth of Knowledge* (Cambridge: Cambridge University Press, 1969).

29. Thomas Kuhn, "Reflections on My Critics," in Lakatos and Musgrave, *Criticism and the Growth of Knowledge,* 231–78.

30. Thomas Kuhn, "Logic of Discovery or Psychology of Research?" in ibid., 1–24 (quote, 21).

31. Karl Popper, "Normal Science and Its Dangers," in ibid., 51–58.

32. Margaret Masterman, "The Nature of a Paradigm," in ibid., 59–90.

33. Kuhn, "Reflections on My Critics," 264. Kuhn would deploy the evolutionary metaphor in a variety of ways. In *Structure* (pp.171–72), he used it to observe that like Darwin's theory his own did not postulate a goal directed process; there was development *from* but not development *toward*. In his "Reflections" essay, he further suggested that each succeeding species had solved increasingly more problems in its environment, just as succeeding paradigms had, and so in both a sense of progress might be retained. The Darwinian analogy became ever more salient for Kuhn, as might be suggested by the tentative title to the manuscript he was working on when he died: "The Plurality of Worlds: An Evolutionary Theory of Scientific Discovery." See Heilbron, "Thomas Samuel Kuhn, 18 July 1922–17 June 1996."

34. Barry Barnes, *Interests and the Growth of Knowledge* (London: Routledge & Kegan Paul, 1977), 23.

35. The actual figures follow. For the plateaus reached in the interval of 1975–1980: of 484 social science journals, 613 articles mentioned *Structure*; of 402 history journals, 232 articles; of 103 philosophy journals, 201 articles; and of 484 biology journals, 53 articles. From 2000 until 2010, the level was about 400 articles in social sciences, 160 in history, 175 in philosophy, and 68 in biology (which thus had a rise in the number of articles mentioning the book). Andrew Abbott, in "Notes on Structure and Sociology" (this volume), has shown that citation to Kuhn's work in humanities, religion, and social sciences climbed linearly until about 1985 and then leveled off at about 450 a year thereafter.

36. The first reader's [Roger Hancock] report (June 22, 1961) notes: "The present ms. is without footnotes or bibliography, or acknowledgements: these would be added according to p. 10." University of Chicago Press Records, Box 278, Folder 4, Special Collections, University of Chicago Library.

37. Thomas S. Kuhn to Sandra Carroll, Copenhagen, March 13, 1963.

38. Thomas S. Kuhn to Carroll G. Bowen, Berkeley, June 29, 1961, University of Chicago Press Records, Box 278, Folder 4, Special Collections, University of Chicago Library.. Kuhn later went back and forth on the advisability of an index in correspondence with the copyeditor (Thomas S. Kuhn to Nancy S. Romoser, Berkeley, May 14, 1962, University of Chicago Press Records, Box 278, Folder 4, Special Collections, University of Chicago Library). He apparently decided against it, and the first two editions appeared without one. The third edition of 1996 was the first to have an index, compiled by Peter J. Riggs.

ONE

Aristotle in the Cold War:
On the Origins of Thomas Kuhn's
The Structure of Scientific Revolutions

GEORGE A. REISCH

The famous opening of *The Structure of Scientific Revolutions* announces the inspiration behind Thomas Kuhn's revolutionary account of science:

> History, if viewed as a repository for more than anecdote or chronology, could produce a decisive transformation in the image of science by which we are now possessed.[1]

For myself, and I suspect many others, this has long been a crucial feature of Kuhn's pioneering work. While he may have gotten some things wrong and gone a little overboard in his talk about "changing worlds," he was right that looking at historical realities, at what he called "research activity itself,"[2] is essential for understanding science.

Lately, I've come to reject this picture of *Structure*'s achievement. Speaking about his historiographic style, Kuhn said he always tried to "get inside the heads" of earlier scientists.[3] But *Structure*'s claims about what occurs in scientists' heads during scientific revolutions appear to be less informed by primary historical sources and more by something inside Kuhn's head, namely, a conception of scientific revolutions that dates to a particular moment in his life.

It occurred in 1947, fifteen years before *Structure* was published. Kuhn called it his "Aristotle experience" and regarded it as crucial for his intellectual development. His publications and papers, including his drafts and outlines for *Structure* as well as his correspondence with his mentor, Harvard president and Manhattan Project administrator James Bryant Conant, confirm that the Aristotle experience was the seed from which *Structure* grew. My claim here is that the more we understand the Aristotle experience and how Kuhn responded to and utilized it, the more we can see how firmly *Structure* and its conception of science was rooted in the distinctive culture of the early Cold War.

One of these roots is an image of what I call "the scientific mind" and, in particular, the mind's relationships to experience and the ideas within it. Some of this mind's properties have been long familiar to readers of *Structure*—it experiences Gestalt shifts in perception and understanding, its observations are "theory laden" and shaped by background beliefs, it accepts new paradigms not on the grounds of strict logical proof, but rather, persuasion.

In the context of the early Cold War, however, a similar image of the mind circulated, albeit in a form highly politicized and, in some cases, militarized. Kuhn's image of the scientific mind whose theoretical commitments and perceptions were transformed when history replaced one paradigm with another was not unlike the mind of American GIs in Korea who were thought to have been "brainwashed" by captors who replaced their liberal ideas and values with those of communism. This was also the mind of the American public that anticommunists such as Senator McCarthy and J. Edgar Hoover sought to protect from communist ideology. And it was the young, impressionable mind of college students, the future leaders of America, whom Conant and other university administrators struggled to shield from communist faculty and the powerful ideology that rendered those professors, as the conventional wisdom then held, unfit to teach.

The Aristotle Experience

Kuhn's most detailed description of his Aristotle experience appears in a lecture from 1981. The experience occurred in the summer of 1947, he recalled, shortly after he began working alongside Conant to develop and teach Natural Science 4, Conant's pioneering attempt to teach science using historical case studies. Being a chemist, Conant asked Kuhn, then finishing his Ph.D. in physics, to develop case studies in the history of physics. That's when it happened.

> I was sitting at my desk with the text of Aristotle's *Physics* open in front of me. . . . Looking up, I gazed abstractedly out the window of my room—the visual image is one I still retain. Suddenly the fragments in my head sorted themselves out in a new way, and fell into place together. My jaw dropped, for all at once Aristotle seemed a very good physicist indeed, but of a sort I'd never dreamed possible.[4]

Kuhn was suddenly convinced that Conant's account of science evolving freely from one conceptual scheme to another, accumulating knowledge and perspective along the way, had to be at least partly wrong. This was the account Kuhn knew from *On Understanding Science*, a book Conant published that year and which Kuhn proofread as its pages came off the Yale University Press. That summer afternoon, however, Kuhn glimpsed something that Conant's image of science could not explain, namely, that Aristotle's physics was not simply mistaken. Instead, it could be understood in a way that made much sense—but only when apprehended along with an array of basic theoretical concepts that were excluded by, and inconsistent with, those presupposed by Newtonian physics. As he would later say of the famous duck-rabbit illusion, Kuhn could see either the Aristotelian duck or the Newtonian rabbit, but not both at the same time.[5]

Structure was conceived at that moment. "Oh, look," Kuhn later told interviewers as he contrasted *Structure* to his first book, *The Copernican Revolution*, which largely follows Conant's picture of science:

> I had wanted to write *The Structure of Scientific Revolutions* ever since the Aristotle experience. That's why I had gotten into history of science—I didn't know quite what it was going to look like, but I knew the noncumulativeness; and I knew about what I took revolutions to be . . . but that was what I really wanted to be doing.[6]

Even as early as 1953, when *The Copernican Revolution* and *Structure* were unfinished, Kuhn contrasted them in a grant application and nodded to *Structure* as "closest to my ultimate purpose as a scholar."[7]

The Aristotle Experience: 1957–1979

An early reference to the Aristotle experience appears in Kuhn's notes for a talk, "A Historian Views the Philosophy of Science," which Kuhn gave at Berkeley in 1957. Eight years earlier, the notes indicate, Kuhn had been a physicist, but "then got invited to assist in a course that involved much H of S. Read it for first time, and particularly read sources."[8]

This, Kuhn exclaimed, was a "shocking experience." His notes read, "Nothing in my physics education or my philosophy reading had prepared me for the way science looks when viewed through writings of dead scientists." The word "shock" appears twice again as he elaborated the differences between science described in textbooks and science revealed in "letters, diaries, laboratory notebooks and, above all, in the articles in scientific periodicals published ten, twenty, thirty years before theory was ready to be embodied in a text."[9]

The shocking experience appears again in a letter Kuhn wrote to Conant in June of 1961. Months before, he had sent Conant a draft of *Structure* with the hope that his mentor would like the book and allow a dedication. But Conant was not much impressed. After some opening pleasantries and encouragement, he firmly criticized Kuhn's overuse of the word "paradigm," his neglect of the "practical arts" and their role in science's history, and his claim that all scientific revolutions involve a change of worldview—a claim that Conant found "too grandiose" and responsible for creating in the manuscript "needless trouble about progress," that is, scientific progress.[10]

At the end of this critique, Conant offered a suggestion, one that Gary Hardcastle has aptly compared to criticism Tennessee Williams might have received about a dress rehearsal for *A Streetcar Named Desire*—along the lines of, "Look, this play has promise. But that character Blanche has just *got to go.*" Indeed, Conant gently encouraged Kuhn to drop paradigms from the book's cast of characters and make do with the familiar language of "conceptual schemes," "climate of opinion," and other such formulations.

The Aristotle experience appears in Kuhn's defensive reply to Conant's charge of "needless trouble about progress." Kuhn wrote:

> You opened <u>On Understanding Science</u> by discussing cumulativeness as the distinguishing feature of science. You then sent me off to look at pre-Newtonian dynamics. I returned from that assignment convinced that science was not cumulative in the most important sense. Newton was not trying to do Aristotle's job better; rather Aristotle had been trying to do a different job and one that Newton did not do so well. Would you say that home industry was *merely* a less effective way of doing what the factory system later did?[11]

Now Steve Fuller has claimed that Kuhn wrote *Structure* effectively at the behest of Conant,[12] but this exchange shows Kuhn rebelling against his mentor and defending his new theory of paradigms. Kuhn's defense continued into the manuscript itself, to which he added another chapter, "The Priority of Paradigms," which elaborates what Kuhn told Conant in this exchange: paradigms are indispensable for understanding science because they are prior to and *more* fundamental than the theories, conceptual schemes, or climates of opinion recognized in extant accounts of how science works.

When *Structure* was published, Kuhn alluded to the Aristotle experience in its introduction. He thanked Conant for exposing him to the history of science and wrote: "to my complete surprise, that exposure to out-of-date scientific theory and practice radically undermined some of my basic conceptions about the nature of science and the reasons for its special success."[13] In a lecture at Michigan State in 1968, "shock" and "surprise" were joined by Kuhn's "astonishment" to discover "that science, when encountered in historical source materials, seemed a very different enterprise from the one implicit in science pedagogy and explicit in standard philosophical accounts of scientific method."[14]

By the time he wrote his preface to *The Essential Tension*, his collection of essays in 1977, Kuhn had begun to describe the Aristotle experience not as the result of extended historical studies in primary sources but rather as a sudden epiphany or "revelation"—a singular moment when the "perplexities" he had always encountered in Aristotle's *Physics* "suddenly vanished."[15] The chronicle within his account now flowed in reverse. What had been a shocking discovery that flowed out of historical study became an epiphany that now preceded his study of history. "Since that decisive episode in 1947," he explained, the "lessons learned while reading Aristotle have also informed my readings of men like Boyle and Newton, Lavoisier and Dalton, or Boltzmann and Planck." The decisive episode, in other words, had itself become a guiding paradigm for Kuhn's future historiography. It revealed "a global sort of change in the way men viewed nature and applied language to it"; it was a revelation, he explained, that informed his "subsequent search for best readings" of historical source material.[16]

The Scientific Mind and Experience

So was Kuhn's Aristotle experience a result of his historical research or was it a guiding intuition about knowledge, about "the way men viewed nature," that preceded it? My view is that the Aristotle experience came first. This shocking experience of 1947 rather quickly led Kuhn to formulate the essential core of the philosophy of science he would debut in *Structure* fifteen years later. *Before* the Aristotle experience, his conception of science was fairly conventional and, indeed, conventionalist with regard to theories.

When it came to the status of "sense experience" or "data," Kuhn was a proud positivist. In an undergraduate essay from the early 1940s titled "The Metaphysical Possibilities of Physics," a young Kuhn was comfortable with the idea that the deep metaphysics of the world, whatever it may be, supported many different and possible theories, all of which he described as "fictions." Yet this young Kuhn was not yet Kuhnian, for these "fictions," he said, were constrained by a stable, independent realm of scientific "data" and "sense impression":

> But while the concepts of physics expand and the narrow fictions are re-placed by broader ones, the structure of physics remains unshaken, for the basis of this structure is data, not concepts, and no change in concepts can invalidate the data.[17]

The Aristotle experience, however, revealed that this was not true. Concepts could, in a way, "invalidate" scientific data. In 1949, shortly after it occurred, Kuhn's papers show him toying with various philosophical and psychological mechanisms that might explain how this 'invalidation' might work. In a sprawling, highly edited, and late-night document that he filed under "Incomplete Memos and Ideas, 1949," he outlined a theory of language, specifically of connotation and denotation that might explain his shocking experience. His guiding idea was that our normal experience of the world was overwhelmingly dense, complex, and filled with epistemic possibilities. We can manage this overflow only because our natural language "cuts" or simplify experience for us:

> What I'm getting at is that natural language provides a finite means of medi-ating an infinitely complex universe. . . . Put more accurately—we in fact live in a world much more complex than our language admits. If we are to act in it, we must simplify it, and our choice of a particular manner of simplifica-tion (a cut) is pragmatically determined and is embodied in our language.[18]

Two systems of physics, in other words, can be consistent and sensible within themselves but utterly different and incompatible side by side if each cuts and reduces the original fullness of experience in mutually in-consistent ways.

The scientific mind as Kuhn was beginning to understand it was *un-aware* of these cuts and simplifications of experience. They were presup-posed and embedded in ordinary language. Two years later, Kuhn slightly modified this picture when describing his current research interests to an administrator in the General Education program. He began again with the overwhelming complexity of a world that "permits an infinity of indepen-dent observations" and required "a choice of those aspects of experience which are to be deemed relevant." But this choice was not made deliberately and consciously by scientists. It was rather made for them by what Kuhn now called an unconscious "predisposition" toward one and only one of the many theories available. "The judgment of relevancy," he explained, "is made on a largely unconscious basis in which commonsense experience and pre-existing scientific theories are intimately intermingled." As a result,

> objective observation is, in an important sense, a contradiction in terms. Any particular set of observations in science (or everyday life) presupposes a predisposition toward a conceptual scheme of a corresponding sort . . . [that] leads the scientist to ignore or discard certain portions of experience in formulating or verifying his theories. But the same 'predisposition' exerts a far more fundamental influence in directing the scientist's attention to particular abstract aspects of experience and blocking his perception of alternate abstractions.[19]

By 1951, then, *Structure's* philosophy of science was largely in place: experience underdetermines theory, theory and observation were dependent and "intermingled," theories were understood as holistic sets of ideas or conceptual schemes, and the scientific mind was unaware that it operates within only one possible system of ideas and that its "perception of alternate abstractions" was "block[ed]"—a lack of awareness that in *Structure* would lead Kuhn to characterize scientific revolutions as "invisible" to most scientists.

This philosophy of science guided the case studies presented in *Structure*, and it plausibly explained the shock and astonishment of the Aristotle experience: until that very moment, the integrity and consistency of Aristotle's physics had been obscured, or "blocked," in Kuhn's mind by Newton's physics and the different cuts and selections from experience on which it rests. Now, Kuhn was becoming Kuhnian and he had effectively taken back his undergraduate proclamation. In his own mind, and in the history of science, Newton's physics had *invalidated* and obscured some empirical data—data that was available to ancients and medievals but not to those who cut the world up along the lines of classical physics.

What was missing at this early stage, of course, was Kuhn's theory of paradigms. But as his difficult exchange with Conant illustrates, paradigms were a late development. Kuhn had used the word before, but in the summer of 1960 he hit on the distinctive formulation that allowed him to finish the monograph he had long before promised Charles Morris, his main contact in Chicago for the *International Encyclopedia of Unified Science*. His new conception of paradigms finally clarified to his satisfaction the "tacit" and nonpropositional character of consensus within a scientific community and allowed him to finish *Structure*.[20] There can be no doubt, however, that paradigms performed the same essential, experience-simplifying roles he had earlier assigned to language and then to "predispositions." During "normal science," a paradigm guided scientists to understand nature in certain ways, and to apply that understanding to only limited regions of experience. As he put it in the chapter "The Route to Normal Science," it guided scientists to "only some special part of the too sizable and inchoate pool of information" that the world presented to us. This guidance, moreover, was largely unconscious: "Scientists work from models acquired through education and through subsequent exposure to the literature often without quite knowing or needing to know what characteristics have given these models the status of community paradigms."[21]

The Captive Scientific Mind

To borrow from another famous Cold War book, Czesław Miłosz's collection of anticommunist essays titled *The Captive Mind* from 1953, Kuhn's normal scientists were mentally held captive by the paradigm under which they worked. Miłosz and Kuhn were both at Berkeley in the late 1950s. My claim, though, is not that Kuhn adopted this image of the mind from Miłosz or from popular anticommunists like Hoover and McCarthy, who routinely appealed to it. For the image of the captive communist mind had taken root in academic culture generally—and at Harvard specifically, where Conant grappled with this national worry about communist faculty precisely during the years he and Kuhn collaborated, from 1947 to 1953.

Conant was no stranger to this notion that certain ideas could captivate the human mind and prevent it from recognizing other ideas and possibilities—in politics or in the history of science. In his *On Understanding Science*, for example, he was fascinated (as Kuhn was later in *Structure*) with the powerful idea of phlogiston and its "almost paralyzing hold on [the] minds" of early chemists. "A chemist reading the papers of the phlogistonists clutches his head in despair," Conant exclaimed, as "he seems to be transported to an Alice-through-the-looking-glass world!" Here, one sees how otherwise very intelligent scientists "twisted and squirmed to accommodate the quantitative facts of calcinations with the phlogiston theory" and remained mired in "a hopeless position."[22]

For anticommunist intellectuals like Conant, Miłosz, and many others, communists were also mired in a hopeless position—inside the all-embracing ideology of Marxism that warped perceptions, impaired judgment, and furnished the true believer with a never-ending supply of rationalizations. Conant's biographer, James Hershberg, has explained that, while the Harvard president had always rejected communism, until 1949 he saw it largely as a mistaken but benign ideology, one that the democratic West could hope to defeat and outlast by cultivating and demonstrating the superior virtues of liberal democracy. Yet that year Conant too experienced a relatively sudden conversion. It was triggered by a cluster of ominous events: Klaus Fuchs, one of Conant's Manhattan Project physicists, was revealed to be a Russian spy; the unexpected news that the Soviets had devised an atomic bomb; and in 1950, the invasion of Southern Korea by Northern Korean and Manchurian forces, a development widely seen as proof that the Soviets were intent on world domination. Before his "conversion to cold warrior," according to Hershberg, Conant had always resisted calls for campus investigations of suspected communist professors; and he had defended academic freedom as an exceptionless principle.[23] But those worrisome events as well as increasing pressure from the Harvard Corporation, from anticommunists in Washington, and from the public (see figure 1.1), led Conant to qualify how far he would go to protect communist faculty. "As long as I am President of the University," the *Harvard Crimson* reported him saying that June of 1949, "I can assure you there will be no policy of inquiry into the political views of the members of the staff and no watching over their activities as citizens." There was, however, a "single exception" to Conant's current thinking on the matter:

> In this period of a cold war, I do not believe the usual rules as to political parties apply to the Communist party. I am convinced that conspiracy and calculated deceit have been and are the characteristic pattern of behavior of regular Communists all over the world. For these reasons, as far as I am concerned, card holding members of the Communist party are out of bounds as members of the teaching profession. I should not want to be a party to the appointment of such a person to a teaching position with tenure in any educational institution.[24]

Communists were "out of bounds," because they had sacrificed their intellectual freedom to their politics. They engaged in "conspiracy and calculated deceit" not by choice or reason but because the dogmatic strings of dialectical materialism, the official philosophy of the Soviet Union, had turned them into puppets of Moscow.

Conant did not create this rationalization. It was largely the invention

Figure 1.1. "Do Colleges *Have* to Hire Red Professors?" The cover
of *American Legion Magazine*, November 1951, illustrating popular concerns
about communist faculty and their subversive influence on young students' minds.

of New York University philosopher Sidney Hook, who tirelessly organized
against communism and left-leaning academics during these years. As can
be seen in figure 1.2, the banner behind Hook indicates that the problem
with communists and fellow traveling sympathizers was that their minds
were not "free." Their political ideas formed a "frontier" beyond which
their natural intellects could not go. As with Kuhn's normal scientists, certain realms of experience were "discarded" and alternative theoretical possibilities were "blocked" from contemplation. That is why, as Hook argued
in his books and in magazine and newspaper articles, communist faculty
were simply unable to fulfill their professional responsibilities as critical,
independent, and creative thinkers.

Hook began to review Conant's books in the *New York Times* in 1948,
and they later met in person at MIT's midcentury convocation in 1949.

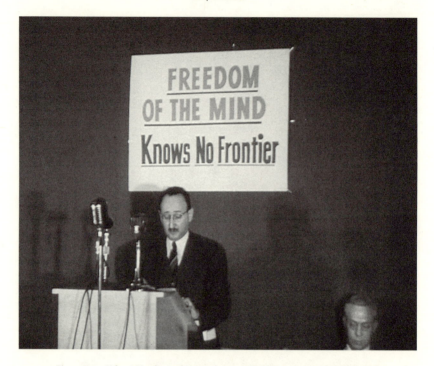

Figure 1.2. Sidney Hook speaking at the counterdemonstration he organized against the Scientific and Cultural Conference for World Peace, New York City, March 1949. (Photo from British Pathé newsreel "Battle of the Pickets," used with permission.)

They subsequently exchanged letters, with Hook usually alerting Conant to some issue, event, or article related to the anticommunist perspective they shared.[25] According to Hershberg, this New York–Cambridge alliance and "the Conant–Hook thesis that Communist Party members were *ipso facto* unqualified to teach" largely set the terms of debate over communist teachers, loyalty oaths, and campus investigations in the 1950s.[26]

I've argued elsewhere that these debates helped transform philosophy of science in North America during the Cold War, largely by making intellectual and social life difficult for philosophers who carried the torch for the unity of science program that was originally promoted by the left wing of the Vienna Circle.[27] In the case of Kuhn and his Aristotle experience, however, America's Cold War preoccupation with mental captivity and sudden ideological conversion was constructive for postwar history and philosophy of science.

On the way toward formulating his mature theory of paradigms, Kuhn held a conception of these "predispositions" that unmistakably nods to the era's fascination with powerful ideologies. In 1953, Kuhn wrote to Morris to propose as the title of his essay "The Structure of Scientific Revolutions" and to explain that its main subject would be "the functions of a theory as a professional ideology for the practicing scientist." Understood as an ideology, theories performed the same two functions, both directing and restricting scientific thought and perception, that he had earlier assigned to language and predispositions and would later assign to paradigms. He explained:

> A theory serves to direct the scientist's attention to certain sorts of problems as "useful" and to certain sorts of measurements as "important;" it dictates preferred techniques of interpretation, and it sets standards of precision in experiment and of rigor in reasoning. Above all, the theory, as ideology, is a source simultaneously of essential direction and of disasterous [sic] inhibition to the creative imagination.

Kuhn even considered using "ideology" in the title of his monograph, but he had second thoughts.[28] Still, he used the word and its variants liberally in notes and outlines for the monograph, as well as in a grant application to the Guggenheim Foundation that he wrote later that year. His forthcoming monograph, he wrote, would be "devoted to the role of established scientific theories as ideologies" that provide "restraints upon the creative imagination."[29]

Paradigms and Conversion Experiences

By the time Kuhn completed *Structure*, of course, "ideologies" had given way to "paradigms" (no doubt to Charles Morris's great relief). But *Structure* and the Aristotle experience that inspired it remain wedded to the surrounding discourse about the possibilities for "ideological conversion" of the captive Cold War mind. Consider the genre of confessional literature that helped many postwar intellectuals, especially former communists, establish their liberal bona fides. The locus classicus of this genre is *The God that Failed: A Confession*, published in 1950, a volume in which Arthur Koestler, Richard Wright, Andre Gide, and other former leftist intellectuals described their growing doubts and then their decisive, and often dramatic, breaks with communism.

Koestler's story is suggestive. Like Kuhn, he was a physicist turned (among other things) historian of science—his *Sleepwalkers* (1959) remains a classic account of Copernicanism alongside Kuhn's first book, *The Coper-nican Revolution* (1957). In *The God that Failed*, Koestler recalled his original conversion to Marxism using much the same language Kuhn would later use to describe his revelation from Aristotle. Koestler's conversion also occurred when he was reading:

> Tired of electrons and wave mechanics, I began for the first time to read Marx, Engels, and Lenin in earnest. By the time I had finished with [Marx's *Theses on*] *Feuerbach* and [Lenin's] *State and Revolution*, something had clicked in my brain which shook me like a mental explosion. To say that one had "seen the light" is a poor description of the mental rapture which only the convert knows (regardless of the faith he has been converted to). The new light seems to pour from all directions across the skull; the whole universe falls into a pattern like the stray pieces of a jigsaw puzzle assembled by magic at one stroke.[30]

Kuhn did not *become* a dogmatic Aristotelian in the same way that Koest-ler became, for many years, a dogmatic Marxist. But the scientists Kuhn portrayed in his case studies of revolutions became dogmatic partisans of a paradigm and spoke of "the 'scales falling from the eyes' or of the 'light-ing flash' that 'inundates' a previously obscure puzzle."[31] Kuhn placed this "parallelism" between scientific and political revolutions in the title of his monograph and defended it in the first few pages of his chapter "The Na-ture and Necessity of Scientific Revolutions." Though he made no attempt to connect these ideological dynamics in science to the ongoing problem of communist professors, Koestler found the connection obvious. It was precisely this exquisite "rapture" of mental captivity that explained how "with eyes to see and brains with which to think," Koestler wrote, there could remain unreconstructed communist intellectuals who "still act in subjective *bona fides*, anno Domini 1949."[32]

In substance, Kuhn's conversion experience—his Aristotle experience— was scholarly and apolitical. But its form was inherited from the sur-rounding preoccupation with ideology and ideological conversion. The intellectual and psychological dynamics—the manner by which one sys-tem of belief and its attendant "cuts" in experience could be replaced wholesale and almost instantaneously with an exclusive alternative—are the same whether the ideas in question are political (with Koestler) or scientific (with Kuhn).[33] Thus, having personally experienced something like an ideological conversion in struggling to understand Aristotle's physics on that hot summer day in 1947, Kuhn would soon glimpse the exciting prospect of a highly original theory of science, one that *embraced* and *normalized* this current image of the human mind as pathologically held captive by its ideas and applied that image to the understanding of science and its history.

Consider again Conant's descriptions of phlogiston chemists struggling irrationally to preserve their beliefs in the face of counterevidence. Inspired by the Aristotle experience, Kuhn realized that in offering descriptions such as those, Conant had picked up the wrong end of the stick. While these particular chemists may have been in a "hopeless position," chemistry itself was going strong—not in spite of, but *because of*, these stubborn, dogmatic phlogistonists who were laying the groundwork for an important revolution in chemistry. That the human mind could be held captive and prevented from exploring other points of view and other regimes of primitive experience—precisely what the Aristotle experience had first revealed to Kuhn—was *necessary* for scientific revolutions to occur and therefore necessary for science's dramatic historical success.

From "Indoctrination" to "Dogmatism"

Was Kuhn aware that his insights about the human mind were deeply rooted in the Cold War culture? His use of the word "ideology," albeit temporarily in the mid 1950s, would suggest that he did,[34] as does an earlier controversy at Harvard over officer training on campus. It began the day after the attack on Pearl Harbor in 1941 when Conant announced to the university community that he would "pledge all of the resources of Harvard University" to the new war effort.[35] In coming months, he pushed forward to militarize the campus for officer training and called on other colleges to do the same. Kuhn was then an undergraduate, but as the controversy escalated he joined the *Crimson's* editorial board and, by July of 1942, had become editorial chairman.

Unlike his predecessor at the helm of the paper, Kuhn consistently approved of Conant's vision for the wartime colleges. Kuhn fully supported Conant in 1942 when the president welcomed the incoming class with a warning that a rigorous education of future military officers by "indoctrination" would be a necessity. This was not the kind of "broadening" liberal education that these freshmen expected.[36] In the paper the next day, Kuhn agreed that this was no mere "compromise of a peacetime ideal" for liberal education. It was rather a "conversion" for war—"a conversion as complete and necessary as that of American industry" that had switched from producing things like refrigerators to bombers. "'Indoctrination' is an ugly word," Kuhn wrote. But "only a plan like President Conant's can ensure the continued use of university facilities when the guns are ready and the college men are called."[37]

At least two elements of the Aristotle experience as Kuhn would later describe it are embedded in this campus controversy: the narrow, purposive cast of the "indoctrinated" mind and the swift "conversion" (Conant used the term "transformation") between these exclusive modes of education and mental styles. Five summers later, when Kuhn was working with Conant and reading Aristotle, he was shocked as these elements suddenly "sorted themselves out in a new way." They had moved from the campus's discussion about military and political discipline into Kuhn's new understanding of how scientific cognition and perception actually work. And their valuation began to shift. "Indoctrination" in particular was no longer so "ugly." For the kind of "specialized education" that created "highly trained manpower"[38] for the war effort had a counterpart, Kuhn realized, in the kind of science education that simplified experience, restricted intellectual creativity, made scientific revolutions "invisible," and thus made "normal science" possible.

Another word no longer so ugly was "dogma." Once *Structure* was finished and sent off to the University of Chicago Press, Kuhn chose this word to introduce the key ideas behind his new theory of science. It is the centerpiece of his essay "The Function of Dogma in Scientific Research" in which he took issue with the popular stereotype of the open-minded (and undoubtedly noncommunist) scientist whose beliefs follow empirical facts and are never enslaved by dogmas. For dogmatism, Kuhn explained, played an essential role in the historical success of science. It was "instrumental in making the sciences the most consistently revolutionary of human activities" and (as he put it in *Structure*) the "drastically restricted vision" of normal science was "essential to the development of science."[39] It made the community sensitive to potential anomalies and helped ensure (as Conant's phlogistonists illustrated so well) that scientists did not abandon their paradigm capriciously or prematurely.

"The Function of Dogma" did not go over very well. Kuhn read the paper just four years after the first Sputnik satellite had convinced Congress and the public that American science education had to be revamped to avoid all traces of dogmatism and authoritarianism. Even though he read the paper in England, far from the anticommunist consensus about American education, the American biologist Bentley Glass, then in charge of the

Sputnik-inspired BSCS reforms in biology textbooks, was present and responded formally to Kuhn's paper with no small amount of indignation. Then and there, he persuaded Kuhn to stop using the word "dogma" and variants when discussing science.[40]

As far as Kuhn's career went, however, this controversial paper may have had a revolutionary and ironic effect. For Kuhn's provocative defense of dogmatism and intellectual conservatism at this British conference destined him to participate in another British conference, the one organized by Imre Lakatos, who chose Kuhn four years later to appear as a foil to Karl Popper and his militantly antidogmatic philosophy of science. That event and the published proceedings, Lakatos and Musgrave's *Criticism and the Growth of Knowledge*, helped to establish *Structure*'s enormous scholarly prestige and influence—even though most essays in the volume were quite hostile. Whether or not the assembled scholars suspected the highly politicized matrix from which Kuhn's central ideas and early terminology were born, this history has remained relatively invisible.

Notes

1. Thomas Kuhn, *The Structure of Scientific Revolutions*, introduction by Ian Hacking, 4th ed. (Chicago: University of Chicago Press, 2012), 1.
2. Ibid., 1.
3. Thomas Kuhn, "A Discussion with Thomas Kuhn," in *The Road since Structure*, ed. James Conant and John Haugeland (Chicago: University of Chicago Press, 2000), 276; see also 280.
4. Kuhn, "What Are Scientific Revolutions?" in *The Road since Structure*, 16. See also Kuhn, preface to *The Essential Tension* (Chicago: University of Chicago Press, 1977) xi–xii.
5. Kuhn, *Structure*, 114; see also Kuhn, "The Relations between the History and Philosophy of Science," in *The Essential Tension*, 6. On reading *On Understanding Science* in proofs, see Kuhn, "A Discussion with Thomas Kuhn," 275.
6. Kuhn, *The Road since Structure*, 292.
7. Kuhn, "Plans for Research," appended to Karl Hufbauer, "From Student of Physics to Historian of Science: T. S. Kuhn's Education and Early Career, 1940–1958," *Physics in Perspective* 14 (2012): 421–70.
8. "Lectures 57–59," Box 3, Folder 10, Thomas S. Kuhn Papers, MC240, MIT Library, Institute Archives and Special Collections. Kuhn's remarks here date the experience to 1949, but others date it to 1947.
9. Ibid.
10. J. B. Conant to T. S. Kuhn, June 5, 1961, p. 3, Box 25, Folder 53, Thomas S. Kuhn Papers.
11. Kuhn to Conant, June 29, 1961, p. 5, Box 25, Folder 53, Thomas S. Kuhn Papers.
12. Steve Fuller, *Thomas Kuhn: A Philosophical History for Our Times* (Chicago: University of Chicago Press, 2000). See, e.g., p. 10: "From this standpoint, *The Structure of Scientific Revolutions* is an exercise in wish fulfillment [on Conant's part]."

13. Kuhn, *Structure*, xxxix.

14. Kuhn, "The Relations between the History and Philosophy of Science," 5, 4.

15. Kuhn, *The Essential Tension*, xi; on "revelation," see Kuhn, foreword to Ludwig Fleck, *Genesis and Development of a Scientific Fact* (Chicago: University of Chicago Press, 1979) vii–viii.

16. Kuhn, *The Essential Tension*, xii, xiii.

17. Kuhn, "The Metaphysical Possibilities of Physics," Box 1, Folder 3, Thomas S. Kuhn Papers.

18. Kuhn, untitled document on meaning, p. 6, Box 1, Folder 6. Thomas S. Kuhn Papers.

19. Kuhn to Professor David Owen, January 6, 1951, Box 3, Folder 10, Thomas S. Kuhn Papers.

20. Kuhn explains this stage of the development of his paradigm theory in his preface to *The Essential Tension*, xix. His use of the word "tacit" suggests a debt he later acknowledged to Michael Polanyi, whose own critique of neopositivism highlighted nonformal but essential aspects of scientific reasoning. On the still-controversial aspects of Kuhn's debt to Polanyi, including Kuhn's attendance at a Polanyi lecture at the Stanford Center in 1960, see Mary Joe Nye, *Michael Polanyi and His Generation* (Chicago: University of Chicago Press, 2011), 242, and Struan Jacobs, "Michael Polanyi and Thomas Kuhn: Priority and Credit," *Tradition and Discovery: The Polanyi Society Periodical* 33, no. 2 (2006/2007): 25–36, and "Thomas Kuhn's Memory," *Intellectual History Review* 19 (2009): 83–101.

21. Kuhn, *Structure*, 17, 46.

22. James B. Conant, *On Understanding Science* (New Haven, CT: Yale University Press, 1947) 80, 86, 89, 95.

23. James Hershberg, *James B. Conant: From Harvard to Hiroshima and the Making of the Nuclear Age* (New York: Knopf, 1993), 483, 493.

24. "Conant Promises No 'Inquiry' Here," *Harvard Crimson*, June 23, 1949.

25. Conant and Hook exchanged at least nine letters between April 27, 1949 and October 16, 1961 (see their papers in the archives at Harvard and at the Hoover Institute, respectively). Hook reviewed Conant's *Education in a Divided World: The Function of the Public Schools in our Unique Society* (New York: Greenwood Press, 1948) in the *New York Times* on October 24, 1948, and he reviewed his book *Education and Liberty* (Cambridge, MA: Harvard University Press, 1953) on February 15, 1953. He mentioned Conant approvingly in his article, "Should Our Schools Study Communism?" (*New York Times*, August 29, 1954) and spoke of Conant as a colleague in his correspondence with others (see Edward S. Schapiro, ed. *Letters of Sidney Hook* [Armonk, NY: M.E. Sharpe, 1995], 220, 222).

26. Hershberg, *James B. Conant*, 609. For a comprehensive analysis of the controversy, see Ellen Schrecker, *No Ivory Tower* (New York: Oxford, 1986).

27. George A. Reisch, *How the Cold War Transformed Philosophy of Science* (Cambridge: Cambridge University Press, 2005).

28. Kuhn to Charles Morris, July 31, 1953, Box 25, Folder 53, Thomas S. Kuhn Papers.

29. Kuhn, "Plans for Research," 458 (original emphasis).

30. Arthur Koestler, "Arthur Koestler," in *The God that Failed: A Confession*, ed. Richard Crossman (New York: Columbia University Press, 2001), 23.

31. Kuhn, *Structure*, 150, 122. On the "parallelism" between scientific and political revolutions, see 92–94.

32. Koestler, "Arthur Koestler."

33. See George Reisch, "The Paranoid Style in American History of Science," *Theoria* 27/3, no. 75 (Sept. 2012): 323–42.

34. The political valencies of *Structure* were also pointed out to Kuhn by his Berkeley colleague Paul Feyerabend, who engaged Kuhn in an extended but apparently fruit-less debate in the spring of 1961 after closely reading a draft of his new book. The circumstances and some of the letters exchanged between Kuhn and Feyerabend are given by Paul Hoyningen-Huene. See his "Two Letters of Paul Feyerabend to Thomas S. Kuhn on a Draft of *The Structure of Scientific Revolutions*," *Studies in the History and Philosophy of Science* 26, no. 3 (1995): 353–87, and "More Letters by Paul Feyerabend to Thomas S. Kuhn on *Proto-Structure*," *Studies in the History and Philosophy of Science* 37 (2006): 610–32.

35. Conant's speech is quoted in Lewis M. Steel, "College Life during World War II Based on Country's Military Needs," *Harvard Crimson*, December 7, 1956.

36. Conant, quoted in "College Days Few, Harvard Is Told," *New York Times*, October 7, 1942. The word "indoctrination" does not appear in those portions of Conant's speech that the *Times* quoted, but Kuhn's quotation of the word implies that Conant used it.

37. Kuhn "Forecast for '47," editorial, *Harvard Crimson*, October 7, 1942. Kuhn's career as *Crimson* editor is described in Hufbauer, "Hufbauer, "From Student of Physics to Historian of Science," 425–26.

38. Kuhn, "Retooling the Colleges," editorial, *Harvard Crimson*, May 13, 1942.

39. Kuhn, "The Function of Dogma in Scientific Research" in *Scientific Change*, ed. A. C. Crombie (New York: Basic Books, 1963), 347–69, (quote, 349); *Structure*, 25. See also Kuhn's 1959 essay "The Essential Tension" in *The Essential Tension*, 225–39.

40. See George Reisch, "When *Structure* Met Sputnik: On the Cold-War Origins of *The Structure of Scientific Revolutions*," in *Science and Technology in the Global Cold War*, edited by Naomi Oreskes and John Krige (Cambridge, MA: MIT Press, 2014), 371–92. On post-Sputnik science-education reforms, see John Rudolph's *Scientists in the Classroom* (New York: Palgrave, 2002).

Kuhn as a young teenager on a camping trip with his father, Samuel L. Kuhn

TWO

A Smoker's Paradigm

M. NORTON WISE

Inhaling

I want to begin my essay by attempting to capture in a very material and sensual way something of the intensity of Thomas Kuhn's presence and his intellectual engagement. The image I am after comes from his unique mode of smoking cigarettes—many cigarettes—in his graduate seminars at Princeton. I participated in several of them between 1971 and 1975, when as a young nuclear physicist I had come to study with him and ended up doing a second Ph.D. in history of science. The subjects already say a lot about his focus: electromagnetic theory, thermodynamics, statistical mechanics, and quantum theory, that is, the foundations of modern theoretical physics. We read only primary sources, many in the original French or German, often quite technical mathematical physics.

Now imagine that there are about eight of us sitting around a seminar table with Tom at the head. The subject is Max Planck's first proof in 1900 of his famous law of black-body radiation, which ultimately would lead to recognition (by Einstein and Ehrenfest in 1906 and Lorentz in 1908) of a fundamental discontinuity of energy, the quantum. Tom has discovered that Planck's proof cannot be assimilated to subsequent canonical proofs and that it involves no discontinuity. Having put a few pertinent equations on the blackboard, he sits down, lights a cigarette, and begins: "Since no such assimilation is possible . . .

[*Intensely concentrated now, looking down at his notes, he begins a truly extraordinary drag on the cigarette, inhaling longer and deeper than one would have thought possible, before continuing.*]

. . . [later analysts] have dismissed his argument as hand-waving."[1] In this remarkable performance at least half the cigarette is reduced to ash, and we all stare in rapt attention, focused on both the cigarette and the argument. This image of Tom at work is inscribed indelibly on my mind. It is an image, not of smoke exhaled into the broad surrounding world, but of smoke followed down into the smallest passages of the lungs.[2]

The incredibly long and deep drag on the cigarette, right in the middle of his sentence, I suggest, embodied a characteristic mental movement of drilling down into the minute detail of an argument. He might then, as in my example, come up with its confusing or dissonant character and the way in which it had been misread by Whiggish scientists or historians looking at it from a later perspective. This is of course the lesson he regularly aimed to convey to his students and readers. It was his starting point for the recognition of anomalies and thereby for the origin of a paradigm change (although he rarely mentioned paradigms). I have used the Planck example not only because it is fairly close to a real instance in the seminar but also because Tom regarded the book in which it appeared, his *Black-Body Theory and the Quantum Discontinuity, 1894–1912* (1978), as epitomizing what he was after in *The Structure of Scientific Revolutions*, as Ian Hacking has also observed in his introduction.[3] It was his last attempt to get a firm historical grip on a paradigm change and to rescue the concept from its undisciplined proliferation of meanings.[4]

A Narrow Technical Thing

With this vignette I have attempted to evoke a sense of how intense was Tom's focus when actually analyzing conceptual shifts. I want to argue that a paradigm, in its most essential role for him as the locus of the production of new knowledge in science, was a very narrow technical thing, the possession of a small group of people who had access to its *precise* and *esoteric* content, accessible to *professional* practitioners of a subspecialty alone. He believed that it was within these confines that the shared paradigm of the subspecialty became productive, as he made particularly clear in the 1969 postscript. "Communities of this sort ["of perhaps 100 members"] are the units that this book has presented as the producers and validators of scientific knowledge. Paradigms are something shared by the members of such groups." Size was important. It even shrank when he reiterated the point a few pages later, distinguishing the view of some readers that he was concerned primarily with "major revolutions such as those associated with Copernicus, Newton, Darwin, or Einstein," from "the rather different impression I have tried to make," of small changes within a small community, "consisting perhaps of fewer than twenty-five people."[5] The terms I have picked out above—*precise, esoteric, professional*—are key words for the role of paradigms within such groups in *Structure*.

Consider first *precise*. A paradigm when it first appears is likely to be rather vague, heuristically valuable but lacking definition. It is during the period of normal science that it acquires precision in its articulation, scope, and relevant measurements. And just because of this precision it can reveal the anomalies that lead to crisis, revolution, and paradigm change. "Anomaly appears only against the background provided by the paradigm. The more *precise* and far-reaching that paradigm is, the more sensitive an indicator it provides of anomaly and hence of an occasion for paradigm change."[6]

The research accomplished through the nuanced use of such a paradigm requires intensive training and practice in order to master the associated equipment, skills, vocabulary, and theoretical articulations. It is *esoteric*. "Acquisition of a paradigm and of the more *esoteric* type of research it permits is a sign of maturity in the development of any given scientific field."[7]

But even in the early days of electrical research, for example, Benjamin Franklin's single fluid theory of the charging of a Leyden jar provided a paradigm that made the transition to maturity possible, suggesting which experiments might be fruitful and which not. Possession of the Franklinian paradigm "encouraged scientists to undertake more *precise, esoteric,* and consuming sorts of work."[8]

In other words, the precision of esoteric paradigms supports what was for Tom the all-important professional specialization of science. *Professional* occurs more than fifty times in *Structure*. It marks the inherently closed character of groups whose communications are largely internal. Unlike Franklin or Darwin, their audience is other specialists. Under these conditions, when individual researchers can take a paradigm for granted, their investigations "will usually appear as brief articles addressed only to *professional* colleagues, the men whose knowledge of a shared paradigm can be assumed and who prove to be the only ones able to read the papers addressed to them." These are the narrow specialists of normal science, which leads to Tom's formulation of the question he posed at the beginning of section III on Normal Science. "What then is the nature of the more *professional and esoteric* research that a group's reception of a single paradigm permits?" Answer: the paradigm enables them to determine significant facts, match facts with theory, and articulate theory, typically all together, thereby producing "a more *precise* paradigm."[9]

Exemplars

My stress on the narrow technical character of paradigms occasions a remark on the exemplars in textbooks that Tom famously associated with
acquiring a paradigm, especially in the postscript, discussed further below.
Certainly he ascribed great importance to the role these shared problem-
solutions and laboratory exercises play for students learning the basic laws
and theories in a discipline. They supply empirical content and implicit
understanding. But I believe the textbook exemplars have been widely mis-
understood to be the immediate referent in normal science for the shifts in
perception that he associated with paradigm change and revolution. This is
the business of professionals, not students.

Although professionals begin to acquire their paradigms through practice in solving the problems set out in textbooks, they necessarily move
on to the much more abstruse problem-solutions that are paradigmatic for
their specialty. Tom was clear about this progression, which preserves the
analogical function of student exemplars while moving them into esoteric
territory:

> That process of learning by finger exercise or by doing continues through-
> out the process of professional initiation. As the student proceeds from his
> freshman course to and through his doctoral dissertation, the problems as-
> signed to him become more complex and less completely precedented. But
> they continue to be closely modeled on previous achievements, as are the
> problems that normally occupy him during his subsequent independent sci-
> entific career.[10]

Textbooks, then, are not the ultimate carriers of the paradigms in which
Tom was centrally interested, shared by twenty-five to a hundred people.
Textbooks are not written for such tiny specialties. Indeed, the most telling exemplars in different subspecialties of a field are so specific that they
signal different paradigms for understanding the most basic theories in the
general field: for example, "though quantum mechanics . . . is a paradigm
for many scientific groups, it is not the same paradigm for them all."[11]

For an example of this specificity I return to the black-body book and
to Tom's "fundamental reinterpretation" of Planck's supposed introduction of the quantum. Planck's referent for past achievement—the paradigm he employed in 1900—was Ludwig Boltzmann's statistical theory of
gases and his attempt to obtain from it the second law of thermodynamics. Planck needed an exemplary problem-solution far more specific than

what was covered in textbooks. Only professional mathematical physicists were conversant with Boltzmann's probabilistic arguments, and even he seldom employed them. Planck was attempting to use Boltzmann's method for establishing the distribution of energy among gas molecules, adapting it for his own model of a black body: a closed cavity containing radiating resonators in the walls in equilibrium with radiation in the cavity. He treated the resonators as analogous to the gas molecules. The problem was to find how the energy was distributed over the resonators/molecules. Like Boltzmann, Planck actually made the calculation by distributing the resonators over the energy, which he divided into cells of small but finite size. But while in Boltzmann's case the size did not matter, and actually disappeared from the result, Planck found that finite size was required for obtaining the correct distribution. That is the result that physicists and historians regularly and Whiggishly misinterpreted as quantization of the resonator energy, limiting resonator energy to an integral number of finite elements. Planck himself made no such interpretation until 1908, after Albert Einstein, Paul Ehrenfest, and Hendrick Lorentz had all drawn that consequence.[12]

What Tom was at pains to make clear was that the difference in interpretation rested on subtly different techniques for counting and different implicit images: distributing finite elements of energy over resonators versus distributing resonators over finite cells of energy. The first image suggested quantization of resonator energy, while the second led only to a puzzle about the partitioning of phase space. This is precisely one of those duck/rabbit shifts in perception that lay at the heart of Tom's conception of paradigm change. The whole narrative, from the esoteric Boltzmann paradigm with its technical details of calculation as exemplar, to the inversion of meanings and radically new conception of quantized resonator energy, to the inability to recognize (even resistance to recognizing) that Planck was not thinking in those terms, is just the story of revolutions that Tom placed at the center of conceptual change in science.

Insulation

Highlighting the Planck example raises what for me has always been the most enigmatic aspect of *Structure*. It is the continual reference to community, in the sense that a paradigm is what a community shares. That identification would seem to invite social and cultural analysis of the community, in terms of both its own structure and how it interrelates with the larger society in which it participates. But Tom's presentation actually allows neither the narrower nor the broader analysis. I will consider first the larger society.

Consistent with common usage in the 1960s Tom employed the internal/external distinction to separate—to *insulate*, in his word—professional scientists from society. Already in the preface he warned readers what they would not find in *Structure*.

> I have said nothing about the role of technological advance or of *external* social, economic, and intellectual conditions in the development of the sciences. . . . Explicit consideration of effects like these would not, I think, modify the main theses developed in this essay, but it would surely add an analytic dimension of first-rate importance for the understanding of scientific advance."[13]

A footnote citing a few of his own historical works that have considered such factors points out again that it is "only with respect to the problems discussed in this essay that I take the role of *external* factors to be minor." The repeated qualifiers apparently reveal a deep ambivalence toward the social. Again further on, while acknowledging that a full account of the Copernican astronomical crisis would include a variety of social and cultural conditions, he nevertheless insisted that such conditions are of secondary significance: "technical breakdown would still remain the core of the crisis. In a mature science . . . *external* factors like those cited above are principally significant in determining the timing of breakdown. . . . Though immensely important, issues of that sort are out of bounds for this essay."[14] Apparently this means that it is only the technical content of paradigms and paradigm change that is roped off from external society.

Why is that? Or rather, what is the function of this internal/external boundary setting? Tom argued that it was necessary in order for the paradigms of professional science to do the work he assigned to them in the puzzle-solving activity of normal science, namely, to set up puzzles with well-defined solutions, the only problems the community would authorize as scientific. "A paradigm can . . . *insulate* the community from those socially important problems that are not reducible to the puzzle form, because they cannot be stated in terms of the conceptual and instrumental tools the paradigm supplies." The same analysis applies to the progressive character of science, for the concentration of the community on the most subtle and esoteric puzzles governs its effectiveness and efficiency. This "very special efficiency" depends crucially on "the unparalleled *insulation* of mature scientific communities from the demands of the laity and of everyday life."[15] Thus, natural scientists are quite unlike engineers, doctors, theologians, lawyers, or, indeed, social scientists, who choose their problems chiefly on the basis of an *external* social need.[16] And this basic difference is intensified by the nearly complete reliance in the education of natural scientists on textbooks, the original source of their skill in using exemplary problem-solutions as analogies, but which exclude social context. In contrast, students in the social sciences maintain contact with the wider world through study of the classical works in their fields and through current research articles.

It is apparent, I think, that any larger sense of the social and the cultural can have no essential role in Tom's scheme for paradigms. It must be eliminated in order for the community to be properly insulated, thereby maintaining its esoteric, professional character, which is the very source of its effectiveness.

Community

What now does *community* stand for and what did Tom understand by introducing what he calls the *sociological* aspect of paradigms in his postscript? I would say that *community* actually extends no further than this or that individual who represents it, who either further articulates and refines its paradigm or challenges it and initiates a paradigm change. Boltzmann, Planck, Lorentz, Ehrenfest, and Einstein are such people for the black-body book, performing the various roles required for the paradigm change from classical-mechanical continuity to quantum-mechanical discontinuity. In analyzing these roles, in digging down into the concrete mathematical work that the classical-mechanical and statistical-mechanical paradigm enabled, Tom's analysis is an extraordinary accomplishment. I admire it immensely. He could not have done it without the intense focus that he brought to all of his work, as well as his seminars, and that I have suggested he displayed in his equally intense manner of reducing an entire cigarette to ash in a single drag, or maybe it was two drags.

Still, what constitutes the community? There are no networks, no exchange relations, no circulation of people or materials, nothing more than the beliefs of individuals, as individuals.[17] And most surprisingly for the black-body case, which has been the subject of some criticism, there are no constitutive interactions with the people who did the measurements on black-body radiation at the national materials testing laboratory in Berlin, the Physikalische-Technische Reichsanstalt. Readers will not discover how a bolometer functioned or why it was perfected there or what it may have had to do with Planck's ideas about resonators and radiation fields. To have opened up such questions would have brought in precisely the kind of external relations that Tom was at pains to exclude from his analysis of paradigms as the essential referent for esoteric professional work. Text and context do not interrelate like foreground and background but are juxtaposed, like the interior of a black body insulated from its surroundings.

This conundrum is only enhanced by the sociological sense of a paradigm that Tom introduced in the postscript. Recall that in order to avoid circularity, he wanted to disentangle the concept of a paradigm from the concept of the scientific community that shares the paradigm. The community was to be identified by *community structure*, including such things as education, membership in scientific societies, journals read, conferences attended, and informal communication networks, without reference to any paradigm the community might share. But research on community structure was not the sort of thing to which Tom devoted any time and it plays no significant role in *Structure*. It is also not the sense of sociological that he was aiming at. That sense occurs in relation to paradigms themselves, when he distinguished two senses of a shared paradigm that occur throughout *Structure*. The first is the more general *sociological* sense, also "more global," while the second is the more limited and technical sense of exemplary problem-solutions, the *exemplars* discussed above. These he calls the deeper of the two, at least philosophically.[18] The sociological sense answers to the question of what the members of a community of specialists share that accounts for "the relative fullness of their professional communication and the relative unanimity of their professional judgment." For this usage he no longer considered the term "paradigm" appropriate and wished instead to recapture the term "theory" broadly construed (which already suggests a problem for the meaning of "sociological"). To avoid confusion he substituted "disciplinary matrix."[19] The matrix has three components (or four, when exemplars are added): "symbolic generalizations," often expressed as laws; "metaphysical" commitments, described as beliefs in particular models; and "values" used in judging theories, like simplicity, consistency, and plausibility, whose application is often strongly dependent on individual personality or biography.[20]

These three components of the matrix, I want to emphasize again, are just the intellectual commitments of individuals, without any reference to group dynamics. There is nothing sociological about them other than that different individuals may agree about them, thereby easing communication; nor of course do they invoke any external relations. The same is true of exemplars, whether considered as a fourth component of the matrix or as a distinct and deeper sense of paradigm, now used appropriately. Exemplars are deeper because, as they become ever more precise and esoteric in specialized research, "they provide the community fine-structure of science,"[21] and because in this role they do the heavy lifting of normal science, revealing the anomalies that lead to scientific revolutions. Once again, we see Tom drilling down into the center of his concerns, pushing aside the distraction of even an extremely weak sense of sociological.

Hostility

I hope now to have shown why Tom's community paradigms not only do not require anything seriously social but actually exclude it externally and transform it internally into intellectual commitments. Social considerations threatened the core significance of a paradigm as productive of new scientific knowledge. All of this was in place before the sociology of scientific knowledge (SSK) got off the ground—before, that is, sociologists who advertised that they were following Tom's lead in *Structure* put forward claims about the social causes of scientific belief that were abhorrent to him. I have another vivid memory of walking innocently by his open office door sometime about 1972 when I heard, "Norton, come in here. Did you see this? Just look at what they're saying," as he stabbed at an article on his desk by a sociologist. Some twenty years later, in his Rothschild Lecture, "The Trouble with the Historical Philosophy of Science," he damned the Edinburgh "strong program" as "absurd: an example of deconstruction gone mad." And he thought the more qualified formulations of others "scarcely more satisfactory."[22] No doubt this public hostility represented his pressing concern that SSK opened up his project (to transform the way we understand science) to renewed charges of relativism, which he had been battling since the publication of *Structure*. But more profoundly in my view, he had been working for years to eliminate social considerations from having any purchase on the exquisitely technical function of paradigms. SSK threatened the heart of his enterprise.

There is a great deal more one could say about Tom's relationship to the much broader social and cultural turn that has come to dominate the history of science, particularly with respect to the large subject of practice, often taken as prior to theory. He remained skeptical of this development until his death, saying more than once in conversation that he could not get practice. I think he meant by this that practice presupposed a paradigm and could not be prior to it, but that is another story. For the moment I want only to reiterate that Tom's skepticism about the social turn was deeply rooted in *Structure*. That makes it all the more ironic that at a recent Kuhn conference in Berlin one sociologically inclined interpreter after another—David Bloor, Harry Collins, Martin Rudwick, and others—celebrated how deeply and directly their perspective derived from their reading of *Structure*.[23]

I can almost see before me the stabbing finger and the deep drag on the cigarette.

Notes

1. I supplement my memory here with a passage from Thomas S. Kuhn, *Black-Body Theory and the Quantum Discontinuity, 1894–1912* (New York: Oxford University Press, 1978), 102.

2. Tom's smoking has featured prominently in the memories of his students. See Jed Z. Buchwald, "A Reminiscence of Thomas Kuhn," *Perspectives on Science* 18 (2010): 279–83, on 279; John Heilbron, as quoted in James A. Marcum, *Thomas Kuhn's Revolution: An Historical Philosophy of Science* (London: Continuum, 2005), 27; Errol Morris, "The Ashtray: The Ultimatum (Part 1)," *New York Times Opinionator Blog*, March 6, 2011, http://opinionator.blogs.nytimes.com/2011/03/06/the-ashtray-the-ultimatum-part-1/.

3. Thomas S. Kuhn, *The Structure of Scientific Revolutions*, 4th ed. (Chicago: University of Chicago Press, 2012), viii.

4. Six years later, he attempted to clarify the ways in which the black-body book "provides the most fully realized illustration of the concept of history of science basic to my historical publications," including paradigms (Thomas S. Kuhn, "Revisiting Planck," *Historical Studies in the Physical Sciences* 14 [1984]: 231–52, on 231).

5. Kuhn, *Structure*, 177, 179–80.

6. Ibid., 65 (emphasis added).

7. Ibid., 11 (emphasis added).

8. Ibid., 18 (emphasis added).

9. Ibid., 20 (emphasis added), 23, 33 (emphasis added).

10. Ibid., 47.

11. Ibid., 49.

12. This sketch summarizes the key chapter 4 of Kuhn, *Black-Body Theory*. See also his "Revisiting Planck."

13. Kuhn, *Structure*, xliv (emphasis added).

14. Ibid., 69 (emphasis added).

15. Ibid., 37 (emphasis added), 164.

16. Ibid., 19, 164.

17. David Kaiser, "Thomas Kuhn and the Psychology of Scientific Revolutions" (chapter 4 of this volume), explores the "psychological individualism" that Kuhn adopted.

18. Confusingly, he twice differentiates these "two very different usages" (Kuhn, *Structure*, 174, 181) and then includes exemplars as a fourth component of the sociological (186) before devoting a separate section to it.

19. Ibid., 181

20. Ibid., 182–86.

21. Ibid., 186.

22. Thomas S. Kuhn, "The Trouble with the Historical Philosophy of Science," Robert and Maurine Rothschild Distinguished Lecture, November 19, 1991 (Cambridge, MA: Department of the History of Science, Harvard University, 1992), 9.

23. Conference at the Max Planck Institute for History of Science, "Towards a History of the History of Science: 50 Years since 'Structure,'" October 17–21, 2012. See also Barry Barnes, *T. S. Kuhn and Social Science* (London: Macmillan, 1982).

THREE

Practice All the Way Down

PETER GALISON

What are the scientific practices behind Thomas Kuhn's philosophy of science, and what do they tell us about that philosophy? We have begun to understand how to think about the science behind the philosophical inquiries of Hermann von Helmholtz, Henri Poincaré, Albert Einstein, and the Vienna Circle—looking at the understanding of science not as a transhistorical analytic framework but instead as unfolding within history, directed toward *particular* forms of scientific work. My aim is to understand the formation of Kuhn's *Structure of Scientific Revolutions*, not to pick at this or that aspect of its use of paradigms or relation to social theory. Instead, the point here is to look at Kuhn's work as a long struggle to assemble on one side what he had learned in civilian (solid-state quantum theory) and wartime (radar countermeasure) research. On the other, using his rather detailed reading and documentary notebooks, drafts, and letters, I want to show how he pieced together a fundamentally *psychological* picture of how the physical sciences functioned, from theory down, not observation up.

Thomas Kuhn's guide into physics was the Harvard theorist John Van Vleck. A student of Edwin Kemble (who wrote the first American "old" quantum theory paper), Van Vleck made the application of quanta to molecular systems his—and an American—specialty. Not for the Americans was the "shark-like" high theory of European atomic physics, philosophical inquiry, advanced mathematics of group theory, and novel formulations of axioms, and matrices. Instead, the Americans learned to calculate—soon producing new and important, *pragmatic* applications of quantum theory in the calculation of the properties of solids as well as spectral lines.[1]

Kuhn grew up intellectually in the aftermath of that first generation of American theorists. In 1963, just months after the publication of his *Structure of Scientific Revolutions*, Kuhn interviewed his former teacher in the context of the oral history of quantum mechanics. Van Vleck commented drily, "As you look back on them, they [early American quantum theorists] were a pretty undistinguished lot. I guess maybe it should be in the record that I was so pleased that there was a reference to Kemble in the first edition of Sommerfeld's that any American physicist should be mentioned was really something."[2] Van Vleck was an outstanding physicist (in 1977 he shared the Nobel Prize), but it is clear, and was clear to Kuhn, that he measured his success against an image of the leading lights of European physics. Was writing up his summary of quantum mechanics in his "Quantum Mechanics and Line Spectra" rewarding work? According to Van Vleck:

> I think so. Because I don't think I would have discovered quantum mechanics. I think it [my review article] served as useful a purpose as, say, another paper or two to flounder around in the old quantum theory. Of course, you can never be sure. In retrospect, of course, I wish I'd followed up and thought more about my correspondence principle for absorption. That was the nearest I ever came to being on the path of the discovery of the true quantum mechanics. I just never had the imagination to follow that up. I might have later, I can't tell, if other things hadn't broken.[3]

Van Vleck, an American pioneer of the new quantum physics, never lost his sense that there was another physics, a more powerful, deeper going, more original science of the quantum that was taking place in Europe under the steering of Werner Heisenberg, Max Born, and Erwin Schrödinger. Bohr's comment about the importance of "non-spectroscopic things" clearly meant a lot to Van Vleck: "I remember Niels Bohr saying that one of the great arguments for quantum mechanics was its success in these non-spectroscopic things such as magnetic and electric susceptibilities."[4] Magnetic susceptibility was how magnetized a substance would become in response to a given magnetic field; electric susceptibility was how electrically polarized something became in a given electric field. Kuhn pursued these quantities theoretically, appropriately enough—Van Vleck had literally written the textbook on the subject. Of his own original contributions to the field, Van Vleck said he had been well positioned for the work, since he had already written a thesis on magnetism for Percy Bridgman and even tried his hand on calculating these quantities using the old quantum theory. When Van Vleck finally reckoned the dielectric constants using the full-on quantum mechanics of 1926, he was one of a (small) crowd able to apply the theory in this domain. "I must confess," Van Vleck told Kuhn, "that [facing simultaneous work elsewhere] rather burned me up because I felt it was a quite significant achievement in quantum theory. When I mentioned it to Bohr he said, 'you should have got me to endorse it, it would have gone through quicker.'"[5]

Van Vleck and the post-1926 generation of American quantum mechanicians were perpetually playing catch up to the Europeans, trying to carve out a domain where they could shine. Van Vleck said, "You always had a little of the feeling that you were one lap behind compared to what was going on in Europe because those people had an inside track of things compared to what we had. I presume you're familiar with Born's MIT lectures. This was the first introduction to the U.S. in English, I would say, of the new matrix mechanics, which I studied very avidly."[6]

This physics world that Kuhn entered in the mid-1940s was ambiguous. Harvard clearly had an outstanding department, but one unquestionably distant from the center of physics that had flourished, and then been destroyed, in Central Europe. Kuhn graduated from Harvard College in 1943, and immediately took up work under the direction of Van Vleck in the Radio Research Laboratory, which had been tasked with developing radar countermeasures to foil German defenses. Again, Kuhn found himself a bit off to the side of center stage—in the theater of radar, the limelight was squarely pointed at MIT. There, and at Columbia University, in hastily assembled buildings and laboratories, the deans of American physics were pushing on a critical war technology, and beyond that, on new instruments, new theories, and new tools. I. I. Rabi was working on radar; he thought it much more critical to the war effort than the atomic bomb. So too was a young generation of experimentalists and theorists: Edward Purcell, Robert Pound, Norman Ramsey, Charles Townes, and Julian Schwinger, just to name a few. In rapid-fire succession, they duplicated and profoundly extended the British work on magnetrons that generated the radar waves; they built up groups to study components of radar, from antennae and receivers to transmitters and display tubes, all while coordinating with military forces on one side and industry on the other. From constructing radars for coastal defense to fire control against attacking aircraft to blind bombing through the European cloud cover, demands for more accurate systems never ceased. These teams of physicists and engineers radically shortened the microwave radar wavelengths down to barely over a single centimeter. The devices they produced made a difference in almost every field of battle.

The Harvard countermeasures team was vastly smaller than the assembled microwave army at MIT—and the Harvard task was much more circumscribed. Their job (they were directed not to communicate with the primary radar-building group so as to avoid a destructive cycle of measure

and countermeasure) was to make jamming devices that would interfere
with German radar. These they did build. But without a doubt the most
successful innovation was decidedly low-tech; the use of aluminum strips,
cut to lengths that would maximally confuse the fire control operators on
the ground, whose antiaircraft guns blasted away at the fleets of Anglo-
American bombers.

Kuhn joined the group and was soon assigned to write theoretical
reports on a variety of topics: what size echoes could be expected from
various ships, how well would the jammers work, and what power levels
would be required to foil a radar signal hitting an Allied ship at distance R
from the enemy radar. Other people would do actual measurements down
at Chesapeake Bay, where the Naval Research Laboratory had a station. Using fairly idealized models of the sea (a flat plane), and a host of empirical and semiempirical parameters, Kuhn listed the reflectivity of a distant
target that had to take into account the effect of the sea ("sea zone") and a
close target that could be treated as if it was in the air ("air zone"). Calculating effects from semiempirical formulae needed to be done—and Kuhn
did it. But it was not, in any sense, a new kind of physics, theoretical or
experimental.

Here is an example: The power density delivered to a radar from a jammer was given by

$$S_j = .3808 \left(\frac{P_j G_j}{R^2}\right) \sin^2\left(\frac{2\pi h_j h_r}{3\lambda R}\right),$$

with S_j in watts per square meter and the other lengths in feet; G_j, the jammer gain; h_j (h_r), the height of the jamming (radar) antenna; R, the distance
of target from the radar; P_j, the power in watts of the jammer pulse, and
λ, the radar wavelength (in feet). As is immediately visible, this was not
a complex and analytically-precise electrodynamic calculation. It was instead rough work ("preliminary, tentative and incomplete" in the author's
words), a workable combination of measured, guessed, and calculated
quantities. Kuhn carefully filled out a table that assembled the measured
radar reflection parameters for a series of ships from the battleship USS
New York and the heavy carrier USS Franklin down to the bow and broad-
side of a surfaced submarine. Using graphical methods to solve some equations, Kuhn extended the work to include a variety of different jamming
devices—all the while recognizing that some of his key parameters could
be off by a factor of ten.[7]

By the time Kuhn arrived in Europe, the flak directed at Allied bombers
by German antiaircraft was falling apart, especially in "blind" conditions.

There were, Kuhn reckoned, five possible factors: personnel, ammunition, and early warning deficiencies on the German side. The Allies' "window" (aluminum, radar-reflecting strips) could be blocking the Nazi radar, or finally the Germans' best radar had not been deployed near front line targets. In the end, Kuhn judged that the German deficiencies were real (according to intelligence). To be safer, nonetheless, he recommended that the U.S. Air Force should put jammers in their B-26s.[8]

From August 25 and 26, 1944: "The candle is too short to attempt a complete account, but I'll try to hit the high spots We landed near the base of the Cherbourg peninsula, just East of the peninsula itself. . . . We passed through St. Lo, an amazing sight the entire town is a shambles. There are scarcely five buildings with roofs or walls and none are unscarred. There seems no reason for people to return to it." Kuhn noted the equipment he came across—like this one on August 27: "Four antenna towers about 150" high. Two unrigged." The retreating Germans had left power supplies, receivers, transmitters, walkie-talkies, telephone lines—Kuhn inspected the sites room by room, recording the abandoned apparatuses by make and frequencies, along with bits of intelligence. "57 Rue Cricourt Alphonse Herzberger—Director of the [Uniprix] Co. He's dead. His wife goes with the Boches."[9] Or two days later: "Interrogation Capt. Kemper (sp.?). . . . There is infrared apparatus. There are 50–60 such machines. . . . They were made in Augsburg by Messerschmidt. Thinks production transferred to Egei. They were airborne." Then the technical questions began, culminating in a summary of procedures: "The approximate position [of the Allied planes] is worked out at the primary center. This is sent to Chateau Beaumont by wire. From their [sic] to Chateau Dutreux from which orders are sent to Luftwaffe in the Hotel Luxemburg by wire." Kuhn learned of the attempts to push the power of the antennae, to change transmission frequencies, to penetrate the radar-blocking fog created by the Allies' "window"—he recorded the Nazi frustration with the lack of bearings needed for the apparatuses and the bugs associated with their most up-to-date receivers.[10]

This was Kuhn's war: calculation of jamming capabilities, liaison between units, working at Harvard, and then hitching rides back and forth just behind the advancing Allied front in France as he and others struggled to piece together the technical and organizational structure of German air defense. It was work in a small unit, often by himself, sending information back up the chain of command. For physicists at the end of the war who had been highly placed in the nuclear weapons laboratories or even the main radar development laboratories, their peacetime work was shaped by a host of radically new forms of work. They had learned how to collaborate with industry and in massive groups, how to garner large-scale government contracts, and how to join federal and university funding. From cyclotrons to newly interdisciplinary teamwork laboratory design, from new forms of computation to a daily collaboration with engineers, the war had taught American physicists a new way of working. So too it was with radar, which had to invent a field (microwave physics) that demanded novel forms of calculation, experimentation, and manufacture. Radio astronomy (in part) emerged from microwave expertise; so did astronomical work (not least Purcell's 21-centimeter work); Nicolaas Bloembergen, Edward Purcell, and Robert Pound's theory of nuclear magnetic resonance; and Charles Townes's maser.

Young Kuhn's work on the counter-radar project was *not* at the microwave frontier, not even close. By the end of the war he was worlds away from the transformed physics that marked short wavelength work at MIT, Columbia, or Bell Laboratories. Kuhn, in a certain sense, had remained within an older kind of work, cocooned, as it were, in small-scale work at the Radio Research Laboratory, working on countermeasures to generate interference in the far-from-cutting-edge German radar. He had a close and good working relationship with Van Vleck, but little knowledge of the advanced mathematical physics developed in the theory group run by Julian Schwinger and his colleagues or by the Harvard experimental physicists. Much later, historian Costas Gavroglu asked Kuhn about his return to civilian physics at the end of the war, "You studied solid-state physics with Van Vleck , . . . Were you interested in the subject itself or in working with Van Vleck?" Kuhn responded, "It was neither. By the time I decided on a thesis topic, I was quite certain that I was not going to take a career in physics. . . . Otherwise I would have shot for a chance to work with Julian Schwinger."[11]

Maybe. It would have been a huge jump from the semiempirical radar countermeasure work or the undergraduate physics courses Kuhn had had to Schwinger's hard-driving, formal, mathematically dense, and unvisualizable quantum field theory. Even Schwinger's wartime Green's function calculations of equivalent circuits had no correspondingly difficult work in the Radar Countermeasures program. (For calibration: Schwinger often insisted, not without a bit of disdain, that Richard Feynman had, with his diagrams, "brought quantum field theory to the masses.") In any event, after the war, Kuhn undertook a thesis problem under Van Vleck's supervision, to which Kuhn contributed a significant new approximation method to calculate certain parameters in what was then called solid state physics: the cohesive energy, the lattice constant, and the compressibility of monovalent metallic solids. This was the kind of problem Van Vleck liked, and in fact, it was, essentially, the theoretical solid state physics (applied quantum mechanics) of America before the war.

You can see Kuhn's prewar physics in his citations as he fought his way to a characterization of metallic properties. Central to his efforts was Eugene Wigner and Fred Seitz's work from 1933–34, where they developed a quantum mechanical method for calculating the properties of metallic lattices. For help with the mathematics, the young physicist (like so many others) referred to Whittaker and Watson's classic textbook *A Course of Modern Analysis* (1927)[12]—again, a long way from the novel methods introduced by Schwinger and others to calculate the properties of radar components. Throughout, Kuhn made good use of Van Vleck's by-then classic text, *The Theory of Electric and Magnetic Susceptibilities* (1932), alongside pre-quantum mechanics work like Bohr's own 1923 work, presented in Harvard physicist Edwin C. Kemble's *Principles of Quantum Mechanics* (1937). You can see graphically the contribution of the Kuhn–Van Vleck correction to what was known from Herbert Fröhlich and Frederick Seitz in the figure 3.1. Note the difference between the dashed and solid line for sodium (Na), potassium (K), and rubidium (Rb), as well as the numerical differences (given in figure 3.2) between Seitz's calculation for lithium, for example. For ε measured in Rydbergs, Seitz had 0.700, and Kuhn, using the standard quantum mechanical (W.K.B.) approximation, had 0.706. For the

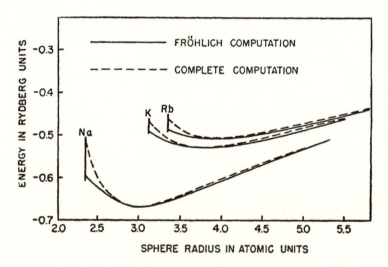

Figure 3.1. Ground-state energy as a function of sphere radius for Na, K, and Rb: comparison of Fröhlich's formula with the complete computations of (A).

TABLE I. Theoretical values of ϵ_{om} and r_{sm}.

Element	ϵ_{om}(Ry.)		r_{sm}(a.u.)	
	W.K.B.	Other[a]	W.K.B.	Other[a]
Li	0.706	0.700	2.84	2.87
Na	0.667	0.668	3.00	2.97
K	0.530	0.530	3.78	3.75
Rb	0.508	0.506	3.92	3.87
Cs	0.468		4.28	

[a] The values for Li are taken from F. Seitz, Phys. Rev. **47**, 400 (1935). The values for the other elements are taken from (**A**).

Figure 3.2. Theoretical values of ϵ_{om} and r_{sm}

radius measured in atomic units, Seitz had 2.87, and Kuhn, with the W.K.B. approximation, got 2.84. A field-changing result this was not.

Still, one should not diminish this work: Kuhn's work with Van Vleck led to an approximation method that was and still is used. But with a few exceptions (Kuhn referred, for example, to the Japanese theoretician, Isao Imai, whose relevant paper appeared in 1948), it was work that, in a very deep sense, was of another, earlier moment in (prewar) history. No fundamentally new instruments were involved in producing the work; Kuhn exploited no radically innovative mathematical or automatic calculation techniques—indeed Kuhn used no novel theoretical concepts. No big collaborations here, no centralized laboratory, no adaptation of war-gleaned knowledge at all. Through both his military and civilian work, Kuhn had seen a physics that was in every way a conglomerate of applications, a working-through of science that had been fundamentally transformed elsewhere. If Van Vleck, despite his remarkable contributions of the 1930s (for which he was awarded the 1977 Nobel Prize), always felt (and said) that American theoretical physics was "a lap behind" the Europeans, Kuhn's was a working-out of an approximation method to Van Vleck's application, a lap behind the lap behind the quantum upheaval of 1926 that had shaken the pillars of physical thought, from causality and visualizability to determinism and locality.

Twice removed from Kuhn's thesis work, shining like distant stars (Kuhn to Van Vleck, Van Vleck to the quantum founders), stood the towering figures of Niels Bohr, Werner Heisenberg, Erwin Schrödinger, and Albert Einstein. By the time Kuhn finished his countermeasure work, he had a strong sense of the productive and necessary function of shared models and textbooks. His own three physics articles squarely built on the published prewar accomplishments of Van Vleck, Seitz, and Wigner; Kuhn cited, and followed, the mathematical physics textbook of Whittaker and Watson, the quantum mechanics text by Kemble, and the defining text on susceptibilities by Van Vleck. Through his war work and the peacetime study of monovalent metals he had gained a real appreciation for the marginal limits of theories with articulated models and better approximations.

Kuhn's whole surround was, to grab his own later term, "normal science": his topic, his working environment, and his techniques—a war, a continent, and twenty years away from one of the extraordinary scientific upheavals in the history of science. In 1949–50, Kuhn was writing up his calculations for publication in the *Physical Review* and the *Quarterly of Applied Mathematics*.[13] Meanwhile, he had been elected to the Harvard Society of Fellows, begun his tenure there, and continued teaching for the university's president, James Conant.

"Notes and Ideas," Kuhn wrote at the top of the first page of a brown "Handy Note Book," and listed for himself his reading through March 1949. Alfred Tarksi's *Introduction to Logic* (check—according to Kuhn's note, that meant "read *in toto*"); Joseph Henry Woodger's *Technique of Theory Construction* (check); Leonard Bloomfield's "Linguistic Aspects of Science" (1939; check); John Dewey's, *Reconstruction in Philosophy* (1920; check); Alfred Jules Ayer, *Language Truth and Logic* (1936; check); and Wiener's *Cybernetics* (1948; check). Somehow, though listed, Immanuel Kant's *Critique of Pure Reason*, John Stuart Mill's *Logic*, and Bertrand Russell's *Scientific Outlook* did not quite get the check of a complete reading.[14]

Right out of the box, Kuhn's position was clear:

> Weaknesses of the positivist or operational position: As Ayer indicates, the doctrine that only questions whose answers can be effected by an activity are meaningful must be taken to be a doctrine of "verification in the weaker sense." Strict verification rules out virtually everything as meaningless. But the notion of "verification in the weaker sense" requires considerable examination."[15]

After ruminating on Ayer's qualifications on "verification in the weaker sense," Kuhn opened a new page, on March 29, 1949, under the header "Language." One should, Kuhn wrote, approach the topic through the writ-

ten language, since it was simpler had fewer signs, and would be more easily formalized than its spoken counterpart. "Some signs (simple or complex) stand for things i.e. are correlated regularly." Then Kuhn broke off his
reasoning and entered in square brackets: "[This is silly: I might as well start
this with a physical vocabulary, say [Rudolf] Carnap's. So let's postpone it.]"
He scratched an enormous "X" through the page. Nothing then until mid-
April 1949, when he came back to Susanne K. Langer (*Philosophy in a New
Key: A Study in the Symbolism of Reason, Rite, and Art*) as well as Robert Mer-
ton's 1939 article "Science, Technology and Society in Seventeenth Century
England" (which he actually read, both got a check) and to Carnap, whose
work Kuhn lists as something he should read, but didn't (or in any case not
sufficiently *in toto* to earn a check). But Piaget he did read, check included,
focusing on the psychologist's 1946 *"Judgment and Reasoning in the Child* as
well as *Notions de vitesse et de movement chez l'enfant."*[16]

Kuhn's reading zeroed in on Piaget from May 21–25, 1949, *Judgment
and Reasoning in the Child* and a few days later and more intensively, *Les
Notions de mouvement et de vitesse chez l'enfant*. On June 14, 1949, he wrote,
"The Piaget reading is useful primarily in shaping my own general view. It
can't be transplanted too literally for the kids haven't got the logical criteria
of the adults I deal with. However . . ." and then Kuhn went on that same
day to list six points that he derived from his reading of Piaget, points I
paraphrase (and excerpt) from Kuhn's notes as follows:

"Egocentrism" (Piaget's term for the earliest of his child development
stages). In the beginning of each science, the emphasis is on the sensu-
ally (observationally) obvious. This is "clearly visible in dealing with the
vacuum notion."

Childish conceptualizations are like those of adults insofar as both begin by
seeing (or potentially seeing) "all," meaning that they can "reconstruct
the experiment correctly." Moreover, the child, like the adult, is capable
of "using conceptualizations (or verbalizations) that ignore or contra-
dict something he 'knows.'" In particular, one can discover the inten-
tion of a word used by the child by noting its conditions of the word's
application—and this way discover the "concept" (Kuhn's term) that the
word connotes. But the child's concepts, though *psychologically* coherent
or satisfying "generally contain logically contradictory elements, that
is, elements such that the word and its opposite can be applied to the
same situation." Only through a "gradual evolution" of experience and
logic are the contradictions "weeded out."[17] Kuhn continued, "Crudely

& metaphorically we may generalize as follows. There is the 'physically visible world' consisting of what we can see with our available sensory & technical equipment. This phys[ical] v[isible] w[orld] is the pure raw flux, unorganized." We structure this flux by creating gestalts or conceptualizations to which we assign symbols and phrases that constitute our "psychologically visible world that 'may contain all or part'" of the physical visible world. Kuhn's layered account starts with the raw input of experience, but only when this input is structured psychologically does it begin to count as objects and their relations: "To 'see' a tree or a velocity difference is to 'see' something in the psych[ological] v[isible] w[world] which is in turn a creation from the phys[ical] real world." Though we may be conscious of the physical world outside psychological one, this is neither necessary nor even predominant, according to Kuhn (reading to the general case through Piaget). Back to the notebook: "Usually when our attention is drawn to these we'll expand the psychological real world to include them." "We are thus conscious of the physic[ical] v[isible] world, but can only 'see' and can only *talk* about the pysch[ological] v[isible] w[orld]."[18]

Children use "élan" for "impetus." Here, Kuhn drew on Piaget's chapter in *Mouvement et vitesse* (which he refers to as M & V) on acceleration.

"Note that in M & V the children are, presumably, capable of recognizing that accelerated motion is not "at uniform vel[ocity]" and yet apply all analytic tools of the uniform velocity. The clue here is that they don't know uniform velocity either tho[ugh] their tools are nearer to fitting it."

"The theory enunciated in (2) is entirely compatible with the importance of 'mental experiments' (as in Galileo, Einstein, etc.) in fact it demands them. Normal scientific method doesn't do this. Also note (M & V, p. 254) the importance of simplifying mental constructions (the extended incline) in the isolation of minimal conceptualization."

Kuhn's final observation is that the role of numbers comes last and incompletely—"note the tendency of [Piaget's] Stage IV kids [so far along developmentally] to use numbers for accelerated motions." The quantification brought to bear on "simple laws" but not to "physical views." The view that quantitative treatment comes late in the scientific process stays with Kuhn through the whole of his work, from these first jottings through his 1962 *Structure of Scientific Revolutions*.[19]

The next day, June 15, 1949, Kuhn came back in his notebook to the analogy between Piaget's children and the history of science. "The outline in 2 above makes phys[ical] science close to s[ocial] s[cience] and to psych[ology] in its method. The difficulties of s[ocial] s[cience] then inhere in the difficulty of recognizing clearly the sources of formal contradictions with the intensions of its vocabulary." Kuhn is saying that all three domains—physical science, social science, and Piagetian child psychology—exhibit moments of "formal contradiction" that must be worked out to enter a new developmental stage. This is straight from Piaget's *Mouvement et Vitesse*, page 263, where, as Kuhn immediately pointed out in his notes, Piaget highlights the mismatch between the psychological and physical world. The child in such a stage registers this point of contradictory beliefs as "groping" (Kuhn's term). On the one side, Piaget notes that the child tries actions that do not correspond to reality; on the other side, he points out that reality corrects the child's expectations. Slowly, with difficulty, expectations and reality begin to align.[20]

Bit by bit, Kuhn attached his account of the history of science to Piaget's childhood-staged sequence of development for speed and movement. At first, the links were hesitant, with cautionary notes—but as he worked through the psychological texts the identification became ever stronger. By the time he concluded his second point in the Piaget comments above, he had left childhood altogether and spoke directly to the adult world of laboratories and blackboards: "The scientific process consists of the attempt to minimize verbal equipment implied by (or inherent in) the psychological r[eal] w[orld]." That scientific process is a continuing one, forged by the encounter of the psychological with the logical and scientific apparatus that brings us the *physical* visible world.[21] Throughout, Kuhn followed Piaget in presenting a bilayer analysis: on the one side there was the physical world and, on the other, its not-always matched representation in the verbal-psychological. Indicating the relative autonomy of the psychological world, Kuhn pointed out, tentatively, there could be other such "worlds" including "the aesthetic & ethical."[22]

This psychologically inflected Kantianism (sharp division between world and representation) then came to an introspective argument: "Everyone has occasionally had the experience of vehemently defending (with complete assurance) an idea or theory which he suddenly, in mid-course, finds totally contradictory. He is then shaken emotionally. Subsequently he can't understand how he could have thought this. Thus we get an illustration of the ease with which one embraces logical contradiction, and the nature of the recanting of ideas upon its discovery."[23] Logic, and for that matter physical reasoning, is decidedly less powerful than the demands of the psychological. "The growth of a sci[entific] conceptualization," Kuhn wrote, "represents a struggle between

a) Pscyh[ological] reasonableness,

b) Logical consistency,

c) Adequacy and applicability to phys[ical] v[isible] w[orld]."

It is, Kuhn continued, not b) or c) but instead psychological reasonableness that is "most important" though "commonly ignored."[24]

With these reflections on the primordial role of the *psychological* view of the world—and the reasoning behind it—Kuhn completed the arc of his first pass toward an account of scientific change. Having begun with the self-admonition that the passage from child to field (ontogeny to phylogeny) shouldn't be taken "too literally," by June 1949, Kuhn had landed that traverse solidly:

> I am supposing that the process Piaget sketches for children takes place at all ages when the range of conceptual thought is extended. This process is of course then redirected to a particular area of thought. . . . New conceptualizations should come from men who were (1) raised in an atmosphere in which these conflicts were implicitly recognized and (2) not set in an old conceptualization. . . . "Scientific contributions must fit the times . . . ," where "fit the times" means [by Kuhn's lights], "be advanced at a time when there's significant intellectual and emotional dissatisfaction with existing 'intensions' to produce the flexibility necessary for acceptance of a new 'intension.'"[25]

This Piagetian psychological "view of the world" dominates for Kuhn—and structures the way he reads Weber. On June 17, 1949, Kuhn, reading the Edward Shils's translation of *Methodology in the Social Sciences*, was ecstatic: "The Weber book is continually brilliant." Kuhn's reading pulled Weber's attention to the *social* sciences over into the *physical* sciences, sometimes explicitly, sometimes not. Take Kuhn's reading of Weber's "'Objectivity' in Social Science." On page 80 of the English translation Kuhn was using, Weber says quite explicitly that he is *not* talking about scientific laws: "not with the 'laws' in the narrower exact natural science sense, but with *adequate* causal relationships expressed in rules. . . . The establishment of such regularities is not the *end* but rather the *means* of knowledge."[26] Now, when Kuhn glossed this section he read the discussion into his notebook not as about *social* laws but as about *scientific laws*. Quoting the above passage, he wrote, "The establishment of such regularities [Kuhn then interpreted "such regularities" as: "i.e. scientific laws"] is not the *end* but instead the *means* of knowledge."[27]

Even more directly yanking Weber into his own psychological idiom, Kuhn, reading Weber's analysis of objectivity, said that Weber's dissection of social scientific objectivity showed a "marked resemblance to my own of physics if the role of [Weber's] 'value' is taken by the 'pscyh[ological] coherence' etc." Here is a final example of Kuhn's reading of Weber. The "ideal type," which, as Weber stresses, does *not* correspond too closely to reality in the social sciences, becomes, for Kuhn, something else: the ideal type is likened to physical idealization—an ideal type is "an aspect of the 'fact.'"[28]

In short, Kuhn's picture was this: a psychological ordering of the world dominates, subordinating both logical and physical orderings of the world around us. When he read Weber, he did so by assimilating it into a fundamentally *psychological*, rather than *social scientific* or more specifically *sociological* frame, and he ventriloquized a Weber who would speak to the physical sciences. With these thoughts about "re-conceptualizations" and a developmental set of steps from observation through articulation to quantification, Kuhn began, on July 5, 1949, to sketch a systematic treatment. "Consider the following outline for a book," Kuhn wrote, calling it *"The Process of Physical Science"*:

Part I. Language and Logic: The Tools of Thought
    Language
    Logic & Math
    New (Linear) Linguistic Modes

Part II. The Scientific Function
    The Physical Real World & the Psychological Real World
    The Problem for Science
    The Emergence of Explicit Tools

Part III. Science at Work. Examples from the History of Science
    Appendix: Relation to other Sciences including Social

The next day, July 6, Kuhn reorganized the book into a simpler, two-part enterprise, putting the psychological up front, pruning the formal linguistic, foregrounding Galileo, and integrating the generalization to other sciences: "The book would be probably easier to write and create less difficulty due to scholarly shortcomings if written as follows":

Part I [Language and Logic: The Tools of Thought]
    The Physical and Psychological World

Language, Logic, and Math
The Adjustment to Science

Part II [The Scientific Function]
1. The Emergence of Explicit Tools and Organized Efforts
2. Cases of Scientific Development & the Impact of G[alileo] G[alilei]
3. The Relation to other Sciences.

Piaget (reinforced by Weber) stands through and through Kuhn's early argument as decisive in his antipositivism. Psychology—not philosophy, not logic, not physical reasoning—was what propelled Kuhn away from the formal positivist philosophy so privileged in textbook versions of scientific theory. Kuhn then explained how a concept (in his first but crucial formulation) could shift its meaning utterly before and after a psychological "re-orientation."

> The development of a word like "velocity" is partly the removal of contradiction in its intension and a shift in the intension itself (Piaget's kids) and partly an increase in the intension itself. This last is accompanied by a switch from v as a transitive quantity to v as a completely intr[ansitive] measure. This doesn't mean we've learned what vel[ocity] *is* but that we've changed the meaning of the word. The shift in meaning shows up as a shift in formal properties.[29]

With these words, Kuhn closed his 1949 notebook.

So it was that by July 1949, Thomas Kuhn had a meaning-centered, developmental, psychologically-driven account of a staged structure of scientific process. Soon Kuhn had a chance to voice his newly configured views, if not to the general public then to a correspondent. Sometime in 1949 (the exact date is unclear), Dr. Sándor Radó, a Hungarian psychoanalyst working at Columbia, wrote to Kuhn, hoping to find support for his view that psychoanalysis could indeed be a "basic science" with experimentation at the center and a sharp transition from "why" to "how" questions marking the triumph of scientificity. The full letters back and forth between Radó and Kuhn have not survived in the archives, but the bulk of Kuhn's tough responses has. Your view of science, Kuhn told the analyst, was a myth, an "immense overestimate of the role of experimentation and of novel observation in the scientific revolution of the sixteenth and seventeenth centuries." This stress on theory over experiment is already a central and often-emphasized feature of Kuhn's thought. He continued, "It is my own opinion that most of the progress in *physical* science before 1750 was achieved by conceptual reorientation toward areas of experience which had been considered by ancient and medieval thinkers, that this reorientation was accomplished without much qualitatively new observation of natural phenomena, that most of the so called crucial experiments of this period were actually designed as demonstrations for sceptics [*sic*] rather than as research tools, and that the *gedanken Experiment* was a more important tool than the physical experiment."[30]

No scientist, Kuhn went on to tell Radó, actually proceeds from the start by producing objective, quantitative work (shades of Piaget). No one. Instead, this process only occurs step-by-step over time, with abstraction playing the truly fundamental role.

> This may be usefully restated in psychological terms by pointing out that observation (and thus experiment) are not capable of the sort of objectivity required for a truly Baconian investigation. At the most elementary level this is shown by psychological experiments on perception, and at a level of greater complexity it is indicated by the experiments associated with the Gestalt school which indicate that we tend to see things first as wholes, that our perceptions of the parts are affected by the manner in which we view the wholes, and that cultural and educational factors may alter our perceptional groupings.

There is, in fact, a failure to see—later generations would go on to wonder how their forebears could be "so dumb."[31]

Gestalts reshaped the parts of an inquiry in sudden and determinative ways. This left Kuhn dubious that "how" and "why" questions could be separated, much less made the basis of scientific status. Any "how" query, argued Kuhn, was only "how" with respect to abstract aspects of the phenomena. Psychology was *always* going to shape what, at a given time, would count as a satisfactory response to "why" questions.[32]

Over the following year or two, Kuhn's views solidified, in part through his reading and research within the open-ended Society of Fellows postdoctoral position, and in part through his experience teaching in the General Education program. On January 6, 1951, Kuhn wrote to historian David Owen, chair of Harvard's General Education Committee, about his field of study and the course he wanted to teach. His emphasis throughout would be on the sources of science, not the end products of research. Such a process rather than product approach could make "an important reorientation in methodological thinking, and I suspect that such reorientation cannot but affect our notions as to the sources of knowledge."[33]

Here in the letter to Owen, Kuhn applied "re-orientation" to the work he was doing—bringing his label of "re-orientation" (applied, most directly to the switch from Aristotle's to Galileo's notion of velocity) to refer to Kuhn's own historical work. (Aristotle to Galileo identified terminologically with the logical positivists to Kuhn.)

> My starting point is that the implicit scientific injunction, "Go ye forth and gather the fruits of objective observation," is a meaningless one which no one could carry out. The complexity of the objects presented by experience permits an infinity of independent observations; so that the process of scientific observation presupposes a choice of those aspects of experience which are to be deemed relevant. But the judgment of relevancy is made on a largely unconscious basis in which commonsense experience and pre-existing scientific theories are intimately intermingled.

Here again, the "psychological view" dominates; as with the early-stage Piagetian child, perception flows in regular, determined channels.

Through insight prized from experimental psychology and linguistic research, Kuhn reported to Owen, he now took it "that objective observation is, in an important sense, a contradiction in terms. Any particular set of observations . . . presupposes a predisposition toward a conceptual scheme of a corresponding sort: the 'facts' of science already contain (in a psychological, not a metaphysical, sense) a portion of the theory from which they will ultimately be deduced." The "conceptual scheme" or "orientation" leads the researcher to attend to some elements and to ignore the rest. In fact, as Kuhn stresses, the conceptual scheme actually *blocks* perception of the (nonconforming) rest.[34]

Still, in the January 1951 letter to Owen, Kuhn brought up again the "re-orientation" of the concept of velocity that marked the last entry of his 1949 notebook. "By the time of Galileo a complete reorientation toward these common experiences had occurred—." Aristotle saw speed as a total distance divided by time, and focuses on the gradual stopping of the pendulum. By contrast, Galileo, saw speed as instantaneous, and it is this fundamental change that is the main step—the quantitative law comes last: "given the reorientation with which Galileo starts, the laws for which he is known could not for long have evaded scientific imagination." In a further development of the outline he had set in 1949, Kuhn now wanted to stress the question: "what reorientation and from what sources," saying he aimed to write a "history of reorientation" in the modern world, for example in the conflict between the mechanical and field theories that led, inter alia, to general relativity and unified field theory.[35]

By the time Kuhn wrote to David Owen, he was deep in preparations to give the Lowell lectures set for March 1951. Above all, Kuhn envisioned his series of talks to begin in history but aim for a philosophical and method-ological goal, the "isolation of certain non-logical, perhaps even psycho-logical characteristics of creative research in physical science."[36] His picture of a physical view and psychological view remained—the latter structuring the flux of the former and it is this two-step that Kuhn had in mind when he assigned a preliminary title to his series: "The Creation of Scientific Objects."

The first of the Lowell lectures (ultimately called "The Quest for Physical Theory: Problems in the Methodology of Scientific Research") commenced on Friday, March 2, 1951. Kuhn laid out his target: the widespread empiri-cist view that science proceeded by dispassionately reasoning from obser-vation to theory. Against this position, he attacked with a very different picture of the work accomplished by Galileo, Dalton, and Lavoisier. De-fending his decision to focus entirely on early science, he argued that first, contemporary science was far too "technical" and "abstract" for the occa-sion. But second and more importantly, Kuhn cautioned his audience that "we" believe in contemporary science, and only older science could offers us the distance needed to study the formation of its conceptual schemes. Nothing would be lost choosing the antique, because "I believe the his-torical unity of science, or more accurately the historical unity of scientists, permits the picture of science which we will derive in this manner to be applied without significant alteration to contemporary science." "Textbook science" is responsible for the widespread empiricist understanding of how science works. Kuhn said it was a "fable," nothing more, to think that our way of justifying science today has anything to do with the creative science that generated it in the first place. In fact, there were "two distinct meanings of the word science." Note the still-sharp imprint of Hans Reichenbach's contexts of discovery and justification. "In the first," said Kuhn, "science is conceived as an activity, as the thing which the scientist does. In its other meaning science is knowledge, a body of laws and of techniques assembled in texts and transmitted from one scientific generation to another."[37]

For Kuhn, only a fabulous account of science could make the past look like its textbook image. About a third of the way through his first lecture, Kuhn wrote in red capital pencil: "BAD HISTORY" (as opposed, presum-ably, to the good history Kuhn was reading in the work of, for example, the

French historian of science Alexandre Koyré). Galileo, so Koyré and Kuhn contended, could not have gotten his results from the Leaning Tower of Pisa—he wasn't there when the mythic history had him doing the experiment; he was just at that moment writing about the physics of fall in a thoroughly incompatible way. Had he, against the facts, tossed wood and lead simultaneously off the tower, he would have seen the lead hit first because of air resistance. Even the inclined plane, Kuhn argued, could not have served as advertised. (Kuhn and his physics colleagues at Harvard had built such an inclined plane apparatus that worked the way it was supposed to, but it took the Cambridge physicists the best modern machine tools and set them back the startling sum of over $500.) In opposition to the textbooks, Kuhn emphasized that Galileo had used "vague facts," "qualitative facts," indeed facts "entirely lacking in numerical precision" in the formulation of his law of acceleration.[38] Theory was no slave to experimental induction, and we would have to look elsewhere for the infrastructure of the "re-orientation" that had taken place between Aristotle to Galileo.

Over the next three Lowell Lectures, Kuhn took on some of the great issues of "early" science: subtle fluids, physical fields, atomism, and dynamics. With a relentless and often gleeful Oedipal bashing of the received "empiricist methodology," Kuhn defended his "homicidal attack" on the Galileo fable and the related fables of Lavoisier and Dalton. Positively, Kuhn used the second half of his lecture series to probe the role that "preconceptions" played in shaping "creative" (not textbook) science. "Can any set of preconceptions prove fruitful? Is the creative scientist actually the man who most strongly displays his individuality of judgment by proceeding from preconceptions different from those of the majority of his profession? And if so what are the sources of these new prejudices? How complete is their domination of research; by what can they be altered?"[39]

Kuhn's responses to his set agenda developed many of the keywords for which Structure would come to be known, starting with the notion of a "crisis." An "orientation," as Kuhn characterized it, functions as a corral of "prejudices and preconceptions," is learned by training, and remains continuous over many years. Already in these 1951 lectures, the orientation is an amalgam of theory and experiment, an "inchoate" combination that can only be replaced by another. This (said Kuhn) is both an accurate historical description and a psychological and logical necessity. Building on the historical examples, Kuhn argued that we can now see the strongly fixed ideas embedded in an orientation can be obstacles to progress, for example, a hard commitment to atomism created an impediment to the development of heat theory or seventeenth-century chemistry. But "orientations" are more than frictional. Views about cosmology (Does the universe have center? Is it infinite or finite? Does the earth move?) open new possibilities for scientific thought. Orientations (or synonymously, "conceptual frameworks") shape similarity relations and fix acceptable analogies. Through such a determination of categories, the orientation shapes the very form of explanation—making it possible, for example, to see circular and linear motion as part of the same "thing." Orientations (returning to Kuhn's original lecture series title) are central to the creation of scientific objects.[40]

But over time, bit by bit, driven by theoretical or empirical difficulties, the old orientation accrues ever more ad hoc assumptions about instruments and objects. Eventually the field enters into a "crisis" stage in which everyone believes in the theory but the theory is so encumbered it begins to sag under its own weight. Eventually (here it is at last), a *scientific revolution* knocks the old orientation out in favor of a new one.[41] Crises could come from economic forces that change motivations, social forces, political sources, speculative philosophy, or changes in cosmology—but this would mean beginning a full-scale sociology of science—which is not where he wants to go here. Such an enterprise would be tying science to, for example, "extra-scientific climate of opinion." Quite deliberately, Kuhn set aside the "extra-scientific," restricting himself to changes in "professional orientation," to shifts of "points of view." (Indeed, on June 14, 1949, Kuhn wrote in his notebook, "Lewis Feuer, *Dialectical Materialism & Soviet Science*"—but not a word about it then or anywhere else in his notes.)[42]

Ending his fifth lecture, Kuhn promised to come back with a more precise "anatomy" of the orientation. To do that, he assigned a "homework problem" which he addressed, ad lib, from the blackboard. Imagine, Kuhn said (as best I can reconstruct his presentation), that you had a square array of alternating types of squares, missing the top right and lower left squares. Could you completely cover all squares by covering in each step two adjacent squares of different types? Kuhn promised his audience a "paradigm" that would be more effective if the problem were stated in advance.[43]

When Kuhn began his sixth lecture at 8 p.m. on Tuesday, March 20, 1951, he offered a restatement of the puzzle: picture the array as a checkerboard and the covering mechanism as a domino that fit over exactly two squares. Now since the opposite corners of a checkerboard are the same color (say, black) and the problem specified that two opposite corners were missing, there are thirty-two red but only thirty black squares. Since a domino must cover one black and one red square, it is immediately apparent that after laying out thirty dominos there will be no black and two red squares left uncovered. But there is no way to cover just two red squares with any number of dominos, so there is simply no way to cover all the squares with two-square dominos. By transforming the problem from something unfamiliar (the array and a rule for covering) to something familiar (a checkerboard and dominos) we experience a reorientation. Said Kuhn: "This puzzle can serve quite successfully as a paradigm of many of the effects of orientation which we have already observed." In particular, the now-obvious fact that the two-square coverings cannot completely cover the array could be put into a long, logical, ad rigorous "textbook form." But the underlying creative insight moved in its own way. The analogy continues: Kuhn pointed out that we could imagine more elaborate rules, extending the dominos to L-shaped blocks covering more than two colored squares. Indeed, we could create a whole new topic in mathematics out of such covering rules.[44]

Now, once a new orientation is in place (say, Galileo's understanding of instantaneous speed), the quantitative presentation of the law becomes imaginable, soluble. But the board-game instance is more or less as far as Kuhn got with the term "paradigm" in the Lowell Lectures. The array-checkerboard problem occupies the role of an exemplar. But it is an exemplar of the general features that Kuhn wanted to point to in the process of "re-orientation." Over the next decade of course, paradigms take over (and develop further) the role that "orientation" or "points of view" played in 1951.

So much here reminds us of *The Structure of Scientific Revolutions* (crises, revolutions, Gestalt switches, allowable analogies, among others) that the differences could be dismissed, wrongly. For Kuhn, there were parallels—deep parallels—that persisted in the Lowell Lectures between psychology of perception and child psychology on one side and science on the other. Kuhn noted that when a subject in a psychology experiment with anomalously marked cards (red spades, for example) sees them in passing, it precipitates a "crisis" of classification. When a Piagetian child-subject gets caught between two conflicting uses of the phrase "as far as," he too enters into a "crisis". Here, on the last page of his script for his sixth lecture, Kuhn handwrote in orange pencil "*SLOW*" (double underlined):

It is because of parallels like this, parallels susceptible of a far more detailed development, that I suggest we equate the notion of scientific orientation with that of a behavioral world. And it is in part the psychological necessity of some behavioral world as a mediator and organize[r] of the totality of perceptual stimuli that I suggest we will never be able to eliminate from the scientific process orientations which originate in experience but which subsequently transcend it and legislate for it.

Kuhn's reformulation of "creative science" had gone far in these lectures—but he was operating entirely within a picture of the world in which a physical world is acted upon and reclassified by a psychological one. The effect may be dramatic on scientists, but it is not, for that, an argument for a multiplicity of worlds, an *ontological* shattering, as is indicated by the last words of the last lecture: "Continuing progress in research can be achieved only with successive linguistic and perceptual re-adaptations which radically and destructively alter the behavior worlds of professional scientists."[45] It is behavior worlds that are destroyed, not worlds full stop.

Applying for a Guggenheim grant for 1954–55, Kuhn reported (probably during the fall of 1953) on his still recent agreement with series editors Charles Morris, Rudolf Carnap, and Philipp Frank that he would write an essay for *The International Encyclopedia of Unified Science*. He would show the vastly more important role that theory plays in scientific development, and the correspondingly limited action of experiment. Theory would direct research, restrain the "creative imagination," restrict the problems deemed by the community "real" or "worthwhile," establish allowable models and metaphors, and dictate the "value judgments" that fix any experimental program.[46]

"Any major shift in the theoretical basis of a science must be *revolutionary* in the destructive as well as in the constructive sense." His future monograph, which Kuhn now titled *The Structure of Scientific Revolutions*, would be the "history of science" contribution to the encyclopedia "devoted to the role of established scientific theories as *ideologies* which direct experimentation and which lend special plausibility to certain sorts of interpretations of experiments. More precisely, I plan to begin by showing that, once established by professional consensus and once embodied in the texts and teaching programs by which a profession is perpetuated, scientific theories play a role far larger than their operationally admissible functions as records of nature's regularity."[47] Experiment demoted, theory promoted.

Over the course of the next six years, there was a slow but systematic swap-out of many of the psychological figures that had figured so large in Kuhn's formulation of his project. Jean Piaget, in my view *the* central figure in Kuhn's early work, the model for stable, coherent, conceptual structures broken by periods of acute disturbance, where meanings become unmoored? Vanished with barely a trace, other than a brief reference in the preface. Heinz Werner? Gone. Max Weber? Not a single reference. In their place appeared an entirely new cast of characters to carry the older tune. For the cognitive and social psychology of perception, there was the work of psychologists Jerome S. Bruner, Leo Postman, John Rodrigues, Harvey Carr, and Albert Hastorf, whose work on gestalts, expectation, memory and observation brought up to date the older work, though it dispensed altogether with the developmental analogy that had been so important for Kuhn.

Famously, Kuhn introduced Ludwig Wittgenstein's notion of family resemblance into his account, though only during the second half of the 1950s. This addition gave him a more philosophically-grounded way of proceeding from exemplary solutions (a signal function of paradigms) to problems solved by students and working scientists. "Conceptual schemes" and "ideologies" went the way of the carrier pigeon. Even "re-orientation," not long before the defining concept of Kuhn's whole project, entered *Structure* exactly once, and even there it deferred to "paradigm change." Kuhn wrote, "One perceptive historian [Herbert Butterfield], viewing a classic case of a science's reorientation by paradigm change, recently described it as picking up the other end of the stick . . . giving [the bundle of data] a different framework."[48] In all these ways, Kuhn moved, incompletely but noticeably, from the structural-developmental psychology of Piaget to a more third-person vision of crises, paradigms, normal science and revolutions.

Despite these shifts, much remained of Kuhn's original formulation—the supremacy of theory, subordination of experiment, and holistic transformations chief among the elements of continuity. This stress on what I have elsewhere called "block periodization" was already apparent to Paul Feyerabend in 1961, while the manuscript was still in prexographic form. "If I understand you correctly," Feyerabend wrote Kuhn, "the ideal is 'normal science' or pattern guided science (science guided by a *single* pattern which everybody accepts with the sole exception of some people you would perhaps be inclined to call cranks). But you never state clearly that this is your ideal . . . you insinuate that this is what *historical research* teaches you. . . . You *falsify* history just as Hegel falsified it in order to finally arrive at the Prussian State."[49] Elsewhere in the letter, Feyerabend hit the theme again: "Your hidden predilection for monism (for *one paradigm*) leads you to a false report of historical event." False, Feyerabend contended, because it ignored the multiplicity of forms of physical reasoning hidden within one of Kuhn's paradigms. "You regard as *one* paradigm (classical physics, for example) which is in fact a bundle of alternatives (contact action: Maxwell vs. action at a distance . . . reversibility . . . vs. irreversibility . . .)." These disputes and conflicts *within* the paradigm of "classical physics" undermined its homogeneity.[50]

Though very obviously not Feyerabend's picture of the ideal form of a more anarchic science of many forms, Kuhn's attachment to the single block went deep, his belief ever solid that outside a revolution the individual researcher was captive to the dominant paradigm—and revolution occurred like a tiny crack propagating, zig by zag, through a solid. Feyerabend wanted more than that, he wanted a willful, driving battle among contenders. "You," he wrote Kuhn, "allow for deviations which are brought about *unintentionally* (deviations, that is, from the original paradigm) whereas you frown up on the *explicit* development of alternatives. What is your *reason* for this position . . . [?] i.e. that alternative to the paradigm which are unintentional side effects . . . are to be welcomed whereas alternatives which are the result of an explicit effort to look for something different are not so good."[51] Feyerabend wanted a scientific street brawl, Kuhn a single Gestalt switch. "All existing philosophies of science," Feyerabend insisted, "(yours included!) are monistic in that they deal with what happens when one paradigm reigns supreme,*" with the asterisk leading to the rebuke: "*You only say that if there are more paradigms, then there will be a mess."[52]

Even Kuhn's invocation of a politically-laden notion of revolution came under fire. "Remember my reservations concerning your comparing *political* revolutions with scientific revolutions. The most fundamental revolution, to me, in the domain of knowledge, would be the transition from a stage of *dogmatism* to a stage where replacement of *any* paradigm is possible . . . Seems to me that political revolutions are more closely related to this fundamental revolution than to changes of paradigms about nature . . ." On this point, Kuhn and Feyerabend would never agree.[53]

Kuhn's achievement in those years from the brown notebook of mid-1949 to *The Structure of Scientific Revolutions* was remarkable. He used his experience as Van Vleck's student and in the Radar Countermeasures group to put scientific practice where logical reconstruction had been. He gave theory its due, not as an auxiliary codification of observations but as a directive, forceful part of what it meant to do physics, or science more generally. He showed the importance of the articulation of an established orientation-ideology-conceptual scheme-paradigm. And he allowed for the startling shock of "revolutionary" work with all the disruption and destruction—as well as production—that attended it.

But we can learn from the real physical, psychological, and philosophical practices that Kuhn had to work within the formative years of his picture of science. His physics—the physics of radar countermeasures and quantum magnetic susceptibilities—was in many ways a vestige of the 1930s. Here was small-scale, mostly individual work. With the quantum revolution now twenty years in the past, the techniques Kuhn needed were mostly available from textbooks like those of Whittaker and Watson, Kemble, Seitz, and Van Vleck. There was no real problem of calculation, no need for the new electronic computers, and no role for simulations carried out by hand or computer. Even Kuhn's war work was mostly individual—whether in Cambridge calculating required jamming power or in France sorting out the lines of communication and authority in the German radar defense.

Just as a generation of work has pried open physics practices, philosophical practices have their own history. Instead of seeing Kuhn's account as a successful or failed gloss of science or physics *in general*, we might do better to see it as a valiant and productive analysis of the physics of the 1930s done in the 1940s about the science of the seventeenth, eighteenth, and nineteenth centuries. Not here do we find the tools to analyze the 2,500 physicists who collaborated to produce the Higgs, nor of wartime Los Alamos, Lawrence Berkeley Laboratory, or CERN. Not in monolithic paradigms and not in the study of textbooks does one find the resources to grapple with the hybrid of algebraic geometry and quantum field theory that forms string theory so present in the last decades of the twentieth century or the first several of the twenty-first. Not in the physics of the late 1930s are analytic or historical tools needed to understand the cross-breed research that is part start-up, part bioprospecting, part global pharma and part biochemistry. Nor is the world of solid state quantum approximations the right place to look at the massively parallel computing power set to work on simulations of galaxy collisions, thermonuclear weapons, or quark plasmas.

Other tools are needed for these other jobs—and we need other means to analyze a world where textbooks and preprints have utterly vanished, giving way to the ArXiv; where the boundaries between disciplines make the biological and physical sciences harder to distinguish; where mathematics and physics, astrophysics and particle physics are in constant, morphing changes. Physics results circulate in new ways, learning is increasingly decentralized, asynchronous, hacked, monetized, and distributed.

And yet: If Kuhn teaches us something about a kind of attentiveness to a theoretically-inflected analysis of the conduct of science; if his works show us the virtue of focusing hard on scientific practices; if it inspires a practice-based analysis of the philosophy of science itself—well, that would be a great and good thing and a legacy worth guarding.

Notes

1. The American physicist Raymond Birge called the European quantum physicists "atomic structure sharks." See the excellent article, Alexi Assmus, "The Americanization of Molecular Physics," *Historical Studies in the Physical and Biological Sciences* 23, no. 1 (1992): 1–34 (quote, 8).

2. John H. Van Vleck, interview by Thomas Kuhn, October 2, 1963, Harvard University, Cambridge, MA. American Institute of Physics website, accessed February 8, 2014, http://www.aip.org/history/ohilist/4930_1.html; John Van Vleck, "Quantum Principles and Line Spectra," *Bulletin of the National Research Council* 10, no. 54 (1926).

3. Ibid. There were, Van Vleck recalled, only a few, mostly younger physicists who were even following the new work:

> Right after the discovery of quantum mechanics, I can remember one prominent mathematical physicist who claimed all the matrix elements were zero. He had a little of an informal public listening to him in the hall. But he had not taken into account the fact that the square of minus 1 need not have the same sign in two different equations, which I pointed out to him. I think you can say that, by and large, only the younger physicists in this country were ever able to get quantum mechanics in their bones. There were exceptions, but, by and large, I think, this was true.

4. Van Vleck, interview by Thomas Kuhn.

5. "As it was, I think Mensing and Pauli beat me to it on being the first to publish that factor one-third. It was essentially a triple tie, though Kronig had it too, all three of us" (ibid.).

6. Ibid.

7. T. S. Kuhn, War Reports, [1943], box 1, folder 8, Thomas Kuhn Papers, MC240, MIT Archives [hereafter, TSK Papers], Cambridge, MA.

8. Kuhn to Commanding General, 9th Bombardment Division, "Desirability of a Carpet Program for Daylight Operations of Medium and Light Bombers," February 7, 1945, box 1, folder 19, TSK Papers.

9. Kuhn, War Notebook, August 27, 1944, box 1, folder 10, TSK Papers.

10. Kuhn, War Notebook, August 29, 1944, box 1, folder 10, TSK Papers. At the end of January 1945, the American British Laboratory (ABL-15) requested that Kuhn, now versed in intelligence aspects of radar countermeasures, be assigned to a liaison position, mediating between the ABL and the 9th U.S. Tactical Air Force.

11. Aristides Baltas, Kostas Gavroglu, and Vassiliki Kindi, "A Discussion with Thomas S. Kuhn," in Kuhn, *The Road Since Structure: Philosophical Essays, 1970–1993*, ed. James Conant and John Haugeland (Chicago: University of Chicago Press, 2000), 274.

12. E. T. Whittaker and G. N. Watson, *A Course of Modern Analysis*, 4th ed. (Cambridge: Cambridge University Press, 1927).

13. T. S. Kuhn, "An Application of the W.K.B. Method to the Cohesive Energy of Monovalent Metals," *Physical Review* 79 (August 1950): 515–19; T. S. Kuhn and J. H. Van Vleck, "A Simplified Method of Computing the Cohesive Energies of Monovalent Metals," *Physical Review* 79 (July 1950): 382–88; and T. S. Kuhn, "A Convenient General Solution of the Confluent Hypergeometric Equation, Analytic and Numerical Development," *Quarterly of Applied Mathematics* 9 (1951): 1–16.

14. Kuhn, Notebook, "Reading to 3–31–49," box 1, folder 7, p. 1, TSK Papers.

15. Kuhn, Notebook, March 29, 1949, box 1, folder 7, p. 3, TSK Papers.

16. On the discussion of language, see ibid., 8. Jean Piaget, *Les Notions de Movement et de Vitesse Chez l'Enfant* (Paris: Presses Universitaires de France, 1946).

17. Kuhn, Notebook, June 14, 1949, box 1, folder 7, pp. 10–12, TSK Papers.

18. Ibid., 14–15.

19. Ibid.

20. Ibid., 18–19.

21. Ibid., 15–16.
22. Ibid., 16.
23. Kuhn, Notebook, June 15, 1949, box 1, folder 7, p. 21, TSK Papers.
24. Ibid., 22.
25. Ibid., 20–21. Kuhn used Heinz Werner's *The Comparative Psychology of Mental Development* (Chicago: Follett Pub. Co., 1948), to bolster his Piagetian conclusions—"synaesthesia shows a general form which psychology can influence perception, from which one can derive primitive notions of "psych[ological] compatibility." So does 'physiognomic perception'" (Kuhn, Notebook, June 15, 1949, box 1, folder 7, pp. 22–23). Kuhn then notes that Werner's first point, on page 128 of *Comparative Psychology*, "is reminiscent of Aristotle's incommensurability of various types of motion" (ibid., 24).
26. Max Weber, *The Methodology of the Social Sciences*, ed. and trans. Edward Shils and Henry A. Finch (New York: The Free Press, 1949), 80 (original emphasis).
27. Kuhn, Notebook, June 15, 1949, box 1, folder 7, p. 24, TSK Papers.
28. Kuhn, Notebook, June 21, 1949, box 1, folder 7, p. 25, TSK Papers.
29. Kuhn, Notebook, July 6, 1949, box 1, folder 7, p. 35, TSK Papers.
30. Kuhn, incomplete response to Sángór Radó, "Incomplete Memos & Ideas," 1949, box 1, folder 6, p. 2, TSK Papers (original emphasis).
31. Ibid., 3–4.
32. Ibid., 6. Kuhn went on to write a fragment called "How Questions, Why Questions, and the Role of Experiment in Physical Science," saying, to be sure, the separation is in textbooks. But: "the prospective worker in a field pretending to the status of science is advised to examine his subject without preconception of embryonic theory, to classify or describe, and only then to generalize. Such a prescription has all the affective values of any appeal for objectivity, but its validity as description and utility as norm may be few or perhaps questioned, since no creative scientific generalizations have been achieved in this manner" (Kuhn, "How Questions, Why Questions, and the Role of Experiment in Physical Sciences," 1949, box 1, folder 6, p. 3, TSK Papers).
33. Kuhn to David Owen, chair, General Education Committee, Harvard University, January 6, 1951, box 5, folder 84, TSK Papers.
34. Ibid.
35. Ibid.
36. The initial invitation to give the lectures is Ralph Lowell to Kuhn, March 8, 1950, box 3, folder 10, TSK Papers. Kuhn responds, accepting, to Ralph Lowell, March 19, 1950, box 3, folder 10, TSK Papers, which includes the quotation given here and the preliminary title. His final title: "The Quest for Physical Theory. Problems in the Methodology of Scientific Research," is contained in Kuhn to William H. Lawrence, April 13, 1950, box 3, folder 10, TSK Papers.
37. The eight unpublished lectures (in hand-corrected typescript form) are located in box 3, folder 11, TSK Papers. "Two reasons to study early science," lecture 1, pp. 7–8; "Two meanings of science," lecture 1, p. 5.
38. Lecture 1, box 3, folder 11, p. 13, TSK Papers; discussion of Galileo, lecture 1, box 3, folder 11, pp. 13–19, TSK Papers. Kuhn cites Koyré explicitly in his manuscript, for example, at lecture 1, p. 20 (Koyré III-23, specifically on Galileo's account that his theoretical considerations should push opponents into confessing the truth of Galileo's view more than experiment). See generally, Alexander Koyré, *Études Galiléennes* (Paris: Hermann, 1939).

39. "Homicidal," lecture 1, box 3, folder 11, p. 21, TSK Papers; Kuhn on "preconceptions," lecture 1, box 3, folder 11, pp. 26ff, TSK Papers. In a later version Kuhn softened this attack and removed the phrase "homicidal attack."

40. For Kuhn's discussion of "Orientations" and "conceptual frameworks" in the Lowell lectures, see lecture 5, box 3, folder 11, , p. 13, TSK Papers; On facts and theories enter together in inchoate form given by orientation, lecture 5, box 3, folder 11, p. 16, TSK Papers [hereafter referring to Kuhn's red pencil paginations]. On orientations being both descriptive and prescriptive: "science has in fact progressed by a series of circular attempts to apply differing orientations or points of view to the natural world. I state this here as a matter of fact, but I think it is also a matter of [psychological and logical] necessity," lecture 5, box 3, folder 11, p. 17, TSK Papers. Kuhn includes a bracketed reference to Emil Myerson as "precursor" to these views about taxonomy and classification, lecture 5, box 3, folder 11, p. 6, TSK Papers; Kuhn includes more on similarity relations, e.g. in lecture 5, box 3, folder 11, p. 13, TSK Papers.

41. On the "crisis stage" of science, see lecture 5, box 3, folder 11, p. 26, TSK Papers; Kuhn comments that only in retrospect are revolutions additive, lecture 5, box 3, folder 11, p. 32, TSK Papers.

42. Kuhn, lecture 5, box 3, folder 11, pp. 43–44, TSK Papers. Kuhn, Notebook, notation on Feuer, June 14, 1949, box 1, folder 7, p. 18, TSK Papers. For a reading of Kuhn and depoliticization, see S. Fuller, *Thomas Kuhn: A Philosophical History for Our Times* (Chicago: University of Chicago Press, 2000); George Reisch, "Anticommunism, the Unity of Science Movement and Kuhn's Structure of Scientific Revolutions," *Social Epistemology* 17 (2003): 271–75.

43. Kuhn, lecture 5, box 3, folder 11, p. 45, TSK Papers.

44. Kuhn, lecture 6, box 3, folder 11, pp. 1–3, TSK Papers.

45. Kuhn, lecture 6, box 3, folder 11pp. 43–45; lecture 8, box 3, folder 11p. 38, TSK Papers.

46. Kuhn, application for Guggenheim for academic year 1954–55, box 5, folder 84, TSK Papers.

47. Ibid., 2 (*revolutionary*, emphasis added; *ideologies*; original emphasis).

48. Citation is to Herbert Butterfield, *The Origins of Modern Science, 1300–1800* (London: G. Bell, 1949), 1–7, iquoted n Kuhn, *Structure of Scientific Revolutions* (Chicago: University of Chicago Press, 1970), 85.

49. The letters from Paul Feyerabend to Kuhn are reproduced in Paul Hoyningen-Heune, "*Two Letters of Paul Feyerabend to Thomas S. Kuhn on a Draft of* The Structure of Scientific Revolutions," *Studies in History and Philosophy of Science* 26 (September 1995): 353–87 (quote, 360; original emphasis).

50. Ibid., 367.

51. Ibid., 372 (original emphasis).

52. Ibid., 375.

53. Ibid., 373 (original emphasis).

Kuhn, left, on vacation with his brother Roger "Rod"
Kuhn and their grandmother Setty Swartz Kuhn

FOUR

Thomas Kuhn and the Psychology of Scientific Revolutions

DAVID KAISER

Looking back in 1969, four years after a rousing session of the International Colloquium in the Philosophy of Science in London devoted to his work, Thomas Kuhn observed wryly, "I am tempted to posit the existence of two Thomas Kuhns. Kuhn$_1$ . . . published in 1962 a book called *The Structure of Scientific Revolutions*. . . . Kuhn$_2$ is the author of another book with the same title." Around the same time he noted that he had been "monitoring conversations" about the book, and "sometimes found it hard to believe that all parties to the discussion had been engaged with the same volume." He lamented to a correspondent that he had begun to "shudder as I discover what my ideas are taken to be." Just as Roland Barthes declared "the death of the author," Kuhn came to realize that his authorial intentions could not dictate how others would read and interpret his work.[1]

Responding to the plasticity of readings inspired by *Structure*, Kuhn made an important change to the book, appending the lengthy 1969 postscript to the second edition. Finding even the postscript insufficient to guide divergent readings, Kuhn composed a series of essays in which he labored to clarify his original intentions, critique particular construals of his work, and highlight areas in which his thinking had evolved since the early 1960s.[2]

In short, Kuhn experienced keenly that his famous *Structure* was no stable baton, passed among an expanding circle of readers; many ran off in their own directions. Behind the well-known essays and published responses to critics, an extensive correspondence survives in Kuhn's papers, stretching over the two decades following the original publication of *Structure*. The collection includes 170 distinct correspondents, with several of whom Kuhn shared multiple, lengthy exchanges. In his unpublished responses to readers, Kuhn crafted and auditioned versions of his later, well-known essays. The letters include anticipations or trial runs of phrasings and disambiguations that appear in the 1969 postscript and in essays like "Second Thoughts on Paradigms," first published in 1974.[3]

Drawing on Kuhn's extensive correspondence, we may chart the dialogic reception of *Structure*. The reception was "dialogic" in at least two senses. Most of Kuhn's responses had an informal tone, akin to an in-person conversation—a tone facilitated, perhaps, by their mode of composition: Kuhn composed most of his letters by speaking into a Dictaphone and then having them transcribed and mailed by an assistant. More important, the reception was dialogic because Kuhn's responses reflected the interests of his interlocutors and the particular points they raised. He did not fashion his responses (or his published essays) in the abstract. Rather, he responded to specific questions and challenges from readers—readers who emphasized certain aspects of his work while ignoring or downplaying others.[4]

Though Kuhn corresponded about his book with scholars from a wide range of disciplines, across the natural sciences, social sciences, and humanities, the largest share came from psychology. Not surprisingly, few of the psychologists pressed Kuhn on such matters as social attributes of learning or institutional frameworks for apprenticeship in the sciences—topics that Kuhn had claimed throughout *Structure* were of crucial importance in the natural sciences, but whose details he had left largely unanalyzed. Rather, most psychologists who wrote to Kuhn focused on matters of individual cognition, or what one correspondent jokingly called "intracranial determinism": the factors within an individual scientist's head that determined how he or she arrives at a particular conclusion.[5]

The psychologists were not responding at random to Kuhn's book. Tracing Kuhn's path to writing *Structure* from the late 1940s to the early 1960s highlights the importance of specific psychological traditions to Kuhn's own evolving thought. Particularly important were the "genetic epistemology" of developmental psychologist Jean Piaget, the "New Look" experimental perception studies of midcentury, and Kuhn's budding interest in psychoanalysis. Kuhn had become fascinated with certain strands of psychological research—bolstered by influential interactions with particular psychologists—as his thinking coalesced toward the first edition of *Structure*. Many of those strands were then amplified, even as they were edited or refined, in his correspondence with psychologists after the book had been published.

Following a brief description of how Kuhn's *Structure of Scientific Revolutions* emerged to become one of the best-known books about science of the twentieth century, I turn to consider the book's composition, focusing on the many legible traces left in the text of Kuhn's own engagements with psychology and psychologists. I then consider Kuhn's correspondence in the wake of the book, before concluding with some suggestions of how situating Kuhn's *Structure* in its web of influences and interlocutors can help make sense of the changing assessments the book has received from historians of science.

Slow Emergence

The Structure of Scientific Revolutions remains unparalleled among works in the history, philosophy, and sociology of science for having reached a mass audience. Cumulative sales exceed one million copies, and at least sixteen foreign-language translations have been published. To mark the fiftieth anniversary of the book's original publication, symposia were held in 2012 at the University of Athens, Boston University, the Max Planck Institute for the History of Science in Berlin, Princeton University, the University of Chicago, and jointly by MIT and Harvard, in addition to a combined plenary session of the History of Science Society and the Philosophy of Science Association at the annual meeting in San Diego. Several journals published special sections to assess the book's legacy in various domains of scholarship.[6]

One would have been hard-pressed to predict such an outpouring of interest when the book first appeared. The University of Chicago Press prepared an initial print run of 3,000 copies (1,000 clothbound and 2,000 paperback), to be released in early September 1962—an average print run for a scholarly monograph at the time. Neither the press nor the book's author anticipated widespread adoption of the book for classroom use, let alone that the book would become a broad-market best seller. Kuhn explained to the press that "it's not a text for any course but might well be assigned, if cheap enough, as collateral reading in phil. of science, etc.," and that he had "no good ideas" about faculty or institutions that might be effective in promoting the book for classroom adoption. He asked the press to send complimentary copies to his parents and several relatives in addition to a small number of professors, several of whom were colleagues at Berkeley, where he was then teaching. A respectable 919 copies sold that first year, and 774 copies the next.[7]

Kuhn and his family were in Copenhagen when the book came out; he had taken a sabbatical from the University of California at Berkeley to head up the massive Archive for the History of Quantum Physics project. Mail sent to him at Berkeley reached him in Copenhagen with delays that varied between weeks and months, so it wasn't until late 1962 that Kuhn began to receive letters in response to *Structure*. One of the first responses to reach him came from Garrett Hardin, a biologist at the University of California, Santa Barbara, who later introduced the notion of the "tragedy of the commons." Hardin enthused to Kuhn that *Structure* was "the best thing that has ever been written" about how science works. Hardin shared a reprint of his own in which he had examined an episode in the history of life sciences, but cautioned Kuhn not to be taken aback by "how primitive my analysis of the historical aspects is." After all, Hardin, conceded, his piece (which had appeared in 1960) had been "written in the year 2 B.K."—before Kuhn. Kuhn was grateful, responding that he had felt "as though my neck were very far out" when publishing *Structure*, "and I have therefore been living with a nervous stomach during the months since the book first appeared." Kuhn repeated that phrase—that he felt as if he had "stuck his neck very far out" in writing the book—in several other letters at the time.[8]

By the mid-1960s, Kuhn had begun to receive a steady flow of correspondence about the book. He heard from professors of history, economics, geography, psychology, physics, and English about their uses of *Structure* in their classroom teaching; the book had earned its place as a textbook after all. A few correspondents even sent Kuhn their students' papers about the book.[9] (For better or worse, the student papers are no longer extant in Kuhn's files; one can only imagine his reaction to receiving dozens of additional student papers to read.) Colleagues like Peter Fox, a historian at the California Institute of Technology, reported on local campus reactions to the book:

> Dear Tom,
>
> I must tell you how interesting, and useful, your 'Structure of Scientific Revolutions' has proved to be. An economist acquaintance at RAND told me about it, and I had the eighteen Freshmen in my European History survey section read it. . . .
>
> The reaction among the older physicists I know here is: 'I haven't read him, but he's all wrong.'
>
> And I notice that in the syllabus of a new linguistics course here, the instructor has listed as item 4 under 'History of linguistic thought,' 'XXth century (on view of Kuhn's paradigms).'
>
> So you are appearing in all sorts of nooks and crannies. I hope your royalties reflect it.[10]

The royalties did indeed begin to accumulate: by the late 1960s, sales had reached forty thousand copies a year.

Not everyone was pleased with the book's repeated appearances on course syllabi. One horrified student at Fairhaven College, a small liberal arts college in Washington State, wrote to Yale physicist and historian of science Derek de Solla Price in 1969 to ask for advice. She had seen that Price had praised Kuhn's book in a review in *American Scientist*, so she inquired whether he thought the book was appropriate for undergraduates "with an extremely limited background in science": "Kuhn's book is being used as a text for six hundred students this year and several more will be using it next year unless we have some evidence that the book is too complex and discouraging [for] students with past histories of poor science courses." Sensing the student's distress, Price responded that he was "horrified that it [Kuhn's *Structure*] is used for liberal arts undergraduates with limited scientific background. Kuhn is a book for professionals, and perhaps its greatest fault is that it has been much misused by non-professionals who find that 'paradigm' is a nice word to roll on the tongue." In place of *Structure*, Price suggested that the students read his own recent book, *Science since Babylon*. Other students wrote directly to Kuhn to express their dissatisfaction (See figure 4.1).[11]

By the late 1960s, the increasing popularity of *Structure* had begun to take a toll. Declining a request in 1967 to lecture on his book at a small college in Wisconsin, Kuhn responded that he had "increasingly and painfully realized that I must drastically restrict my activities as a propagandist for my views if I am ever to get the time to write another book." Early in 1970 he reported to another correspondent that he had "been getting perhaps a letter a week about my book ever since it appeared almost seven years ago." Many of those letters included manuscripts on which the senders hoped Kuhn would comment; Kuhn lamented the combined "intellectual" and "diplomatic" problems of coping with so many unsolicited materials. By the summer of 1973, he had resorted to a kind of form letter. Several letters went out that season with identical wording:

> For better than five years I have been receiving two or three unsolicited manuscripts, sometimes of book length, every week. Though I regularly mean to write more than routine acknowledgments, I find that, in fact, very few of them do get read. Though I very much hoped that my *Structure* would be widely read, I never dreamed of the nature or magnitude of the problems which its success would create for me.

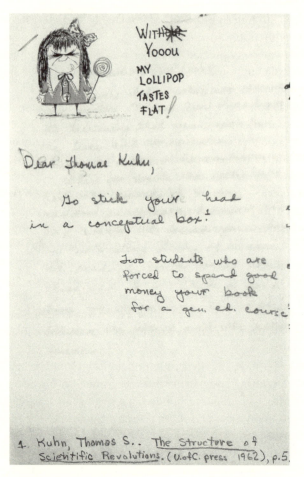

Figure 4.1. Unsigned letter to Thomas Kuhn, ca. January 1968.
(Thomas S. Kuhn Papers, Institute Archives, Massachusetts Institute
of Technology, folder 4:14.)

At the same time Kuhn was feeling inundated, his book made even sharper
inroads into the academy. Between 1976 and 1983, *Structure* was the most-
cited book in the *Arts & Humanities Citation Index*, cited more often than
works by Freud, Wittgenstein, Chomsky, Derrida, or Foucault.[12]

Composition

Though *Structure* first appeared in 1962, Kuhn often recalled that he had
begun thinking about the material fully fifteen years earlier. While pursu-

ing his Ph.D. in theoretical physics at Harvard in the late 1940s—where he studied solid-state physics with Nobel laureate John Van Vleck—he also began teaching in Harvard's then-new General Education program, working closely with Harvard's President James B. Conant. Writing a few years later, Van Vleck noted how disappointed he was that Kuhn began spending so much time in General Education. "He would have stood up very well in theoretical research in quantum mechanics, had he wished to make this his main pursuit," Van Vleck observed.[13]

In 1947, while helping to prepare a new General Education course, Natural Sciences 4, Kuhn had what he later described as his "Aristotle experience." He had been reading Aristotle's descriptions of motion and, with a Newtonian framework in mind, he found Aristotle's account "full of egregious errors, both of logic and of observation." But he kept reading:

> I was sitting at my desk with the text of Aristotle's *Physics* open in front of me and with a four-colored pencil in my hand. Looking up, I gazed abstractedly out the window of my room—the visual image is one I still retain. Suddenly the fragments in my head sorted themselves out in a new way, and fell into place together. My jaw dropped, for all at once Aristotle seemed a very good physicist indeed, but of a sort I'd never dreamed possible. Now I could understand why he had said what he'd said, and what his authority had been. Statements that had previously seemed egregious mistakes, now seemed at worst near misses within a powerful and generally successful tradition.

Kuhn often narrated this experience as akin to a Gestalt switch. It became the prototypical version of his famous description of revolutionary changes in science, "in which some part of the flux of experience sorts itself out differently and displays patterns that were not visible before."[14]

Soon after his "Aristotle experience," just as he was wrapping up his physics dissertation, Kuhn began a three-year fellowship in the prestigious Harvard Society of Fellows (1948–1951). His reading notes from that period survive. Alongside major works from the Western philosophical canon, he also began to immerse himself in writings by developmental psychologists Jean Piaget and Heinz Werner.[15] Kuhn read in the opening chapter of Piaget's *The Child's Conception of Physical Causality* (1930), for example, that young children describe physical motion in distinctly Aristotelian terms, explaining that a projectile continues to move after being thrown because "the air pushes it." Piaget argued throughout his book that children pass through distinct stages in their reasoning about physical motion. The stages mirrored those of the history of science, from the ancient

Greeks through more modern notions of impetus and inertia—a mental version of Ernst Haeckel's famous dictum that ontogeny recapitulates phylogeny. In his conclusion, Piaget posed the central question that would occupy Kuhn's own wrestling with positivism and antipositivism: "If the child's mind is active in the process of knowing, how is the collaboration effected between his thought and the data of the external world?"[16]

Piaget expanded on this theme throughout *The Child's Conception of Movement and Speed* (1946), another book that Kuhn read while in the Society of Fellows. Piaget agreed with phenomenologists and Gestaltists about the "radical interdependence" of knowing subjects' preexisting ideas and how they ordered their sense impressions. "It is clear that the same empirical observation," he wrote, "will give rise to quite a different construction according to the way it is understood"—an anticipation of what Kuhn, Paul Feyerabend, and others would later dub "theory-laden observations." To Piaget, it was not enough to highlight the role of preexisting ideas. One had to consider how people's "schematization" changed over time: "This understanding is transformed in terms of the mental stages already traversed." Moreover, the act of imposing order on sense impressions, of seeing-through-a-schema, constitutes "a modification of reality by the subject." Or, as Kuhn would later write, when Lavoisier began to see oxygen while Priestley saw dephlogisticated air, Lavoisier "worked in a different world."[17]

It was around the same time that Kuhn encountered Heinz Werner's writings. Werner, like Piaget a developmental psychologist, had developed an experimental program devoted to perception while director of the Psychology Department at the University of Hamburg in the late 1920s. A colleague of art historian Erwin Panofsky, Werner had also been closely associated with the Leipzig school of Gestalt psychologists before he was forced to flee Germany when the Nazis took power in 1933. Settling in the United States, he expanded an earlier study into a massive English-language tome, *Comparative Psychology of Mental Development*, which went through multiple editions in several languages. For the 1948 edition, Harvard psychologist Gordon Allport added a foreword, praising the book for making German traditions of Gestalt psychology more accessible to American audiences. Allport praised Werner's book soon after he (Allport) had helped to establish the new Department of Social Relations at Harvard; perhaps the local ferment helped bring Werner's book to Kuhn's attention.[18]

Like Piaget, Werner emphasized various "structures," or levels, of thinking through which individuals and entire societies pass. Even more than Piaget, Werner argued that development between stages was highly nonlinear and noncumulative: "Each higher level is fundamentally an innovation, and cannot be gained merely by adding certain characteristics to those determining the preceding level." Rather than "unbroken continuity," Werner concluded, changes between levels are punctuated by "crises."[19]

Reading Piaget and Werner clearly left their mark on young Kuhn. He briefly mentioned encountering works by Piaget and the Gestalt psychologists in the preface to *Structure*, but other interactions suggest even stronger influences. His wife at the time, Kay, recalled that Piaget visited the Berkeley area to deliver some lectures in the late 1950s, while Kuhn was on the faculty there. Kuhn felt shy about approaching the famous psychologist even though he had been so inspired by his work. Kay commanded: if Kuhn didn't introduce himself to Piaget, then he shouldn't bother coming back home! Meanwhile, as late as 1965, Kuhn recommended the 1948 edition of Werner's *Comparative Psychology of Mental Development* to a correspondent who had asked for reading suggestions about the psychology of perception.[20]

Kuhn's encounters with developmental psychology were rather bookish. He interacted more directly with acolytes of the "New Look" school of experimental psychology. Jerome Bruner and Leo Postman's work became especially important. Like Kuhn, both Bruner and Postman had completed their Ph.D.s at Harvard in the 1940s, and both were young professors in the new Department of Social Relations while Kuhn was in the Society of Fellows. In 1949 they published their paper, "On the Perception of Incongruity: A Paradigm." More than the word "paradigm" impressed Kuhn; he would later devote several pages of *Structure* to summarizing their findings. Bruner and Postman measured how much time test subjects required to identify playing cards, and found that subjects took longer to identify cards that were intentionally "incongruous" (or, as Kuhn would later say, "anomalous"), such as a red five of clubs or a black ten of hearts. To Bruner and Postman—and to Kuhn—the study highlighted how much a person's preexisting ideas or expectations could impact seemingly "direct" empirical observation.[21]

Following his three years in the Society of Fellows, Kuhn began teaching full time in the General Education program at Harvard. As early as 1951, he began to incorporate his various psychological encounters during his society fellowship in his thinking about science. In his Lowell Lectures that year, he emphasized Gestalt-like switches among conceptual orderings of the world and the roles that psychological "predispositions" play in scientists' observations. Two years later he was invited to contribute the entry on "history of science" for the *International Encyclopedia of Unified Science*. The commission—likely an outgrowth of his participation in the "Unity of

Science" discussion group organized by Harvard physicist and philosopher Philipp Frank—set Kuhn on the path to writing *Structure*.[22]

In 1956, Kuhn left Harvard to take up a joint appointment in history and philosophy at Berkeley. In addition to bringing him into close contact with philosophers Paul Feyerabend and Stanley Cavell, the move also solidified his connections with psychologists. Leo Postman, for example, had left Harvard for Berkeley in 1950, and it appears that he and Kuhn reconnected after Kuhn arrived. When describing Bruner and Postman's famous experiment with the playing cards in *Structure*, Kuhn reported, "My colleague Postman tells me that, though knowing all about the apparatus and display in advance, he nevertheless found looking at the incongruous cards acutely uncomfortable."[23] Kuhn spent the 1958–59 academic year at the nearby Center for Advanced Study in Behavioral Sciences at Stanford University. He later recalled that year as a pivotal one in his writing of *Structure*—an early draft of the book from that time survives in his papers—and while there he befriended James Jenkins, a cognitive psychologist and psycholinguist from the University of Minnesota who was also spending a sabbatical year at the Stanford center. A few years after their stint, Jenkins reminisced with Kuhn about "the good old days" they had shared there; the two remained in touch for years afterward.[24]

During that same time, Kuhn began to meet with psychiatrists and psychoanalysts in the San Francisco area. As Kuhn later reported to a psychiatrist acquaintance, he had "given one seminar to a part of the group, and . . . had numerous other opportunities to explore the extent and nature of their interest" in the themes that would eventually appear in *Structure*. "All of this has been good for my intellect and my morale."[25] He had read Freud's *Psychopathology of Everyday Life* as a young man and had also undergone psychoanalysis during the 1940s. Later in life he seemed uncertain whether psychoanalysis had much therapeutic power (at least for himself), but "it sure as hell is interesting." In his discussions with psychiatrists and psychoanalysts, he drew a distinction between "psychoanalysis as therapy and psychoanalysis as science." The latter, it seems, offered Kuhn one more resource with which to think about the active role played by an individual's mind in construing the facts of experience: the "same" situation could be interpreted differently by individuals in distinct mental states, pointing (as much as the "New Look" perception studies did) to the priority of mental particulars over seemingly empirical evidence from the outside world.[26]

With these psychological influences and relationships in mind, we may consider Kuhn's rhetorical strategies throughout *Structure*. Though the book was commissioned as an exercise in the history of science, virtually every important philosophical claim in the book—antipositivism and theory-laden observations, radical ruptures between periods of normal science, incommensurability and the notion of "different worlds" for inhabitants of successive paradigms—is motivated by analogy to experimental psychology, rather than defended by close historical analysis of primary sources. In section 6 on anomalies and perception, for example, Kuhn wrote of the Bruner-Postman experiment:

> Either as a metaphor or because it reflects the nature of the mind, that psychological experiment provides a wonderfully simple and cogent schema for the process of scientific discovery. In science, as in the playing card experiment, novelty emerges only with difficulty, manifested by resistance, against a background provided by expectation. Initially, only the anticipated and usual are experienced even under circumstances where anomaly is later to be observed.[27]

In section 8, on scientists' responses to crises, Kuhn invoked "the switch of a gestalt, particularly because it is today so familiar," and hence can serve as "a useful prototype for what occurs in full-scale paradigm shift," even if the implications of such ruptures for understanding the practice of science "demand the competence of the psychologist even more than that of the historian."[28]

To counter the positivists' long quest for a "pure observation-language," Kuhn wrote in section 10: "modern psychological experimentation is rapidly proliferating phenomena with which that theory can scarcely deal." Experiments with the duck-rabbit gestalt image demonstrate that people with "the same retinal impressions can see different things," while experiments with inverting lenses demonstrate that people with "different retinal impressions can see the same thing." Indeed, Kuhn concluded, "Psychology supplies a great deal of other evidence to the same effect"—evidence of the sort that filled Werner's 1948 book.[29] In that same section, Kuhn drew on the notion of schemas from developmental psychologists like Piaget and Werner to argue that

> the child who transfers the word "mama" from all humans to all females and then to his mother is not just learning what "mama" means or who his mother is. Simultaneously he is learning some of the differences between males and females as well as something about the ways in which all but one female will behave toward him. His reactions, expectations, and beliefs—indeed, much of his perceived world—change accordingly.

The same process occurs, argued Kuhn, as scientists learn to apply concepts to make sense of the world holistically rather than "piecemeal or item by item."[30]

Residues from Kuhn's encounters with specific psychological traditions thus run throughout *Structure*. Developmentalists' focus on children's learning processes, with their nonlinear, noncumulative stages, provided a framework with which to make sense of Kuhn's own "Aristotle experience." The demonstrations by "New Look" perception psychologists that predispositions affect individuals' perception and cognition bolstered Kuhn's antipositivist notions of theory-laden observations and the roles of anomalies. And his interest in psychoanalysis further focused his attention on individuals' active construal of the givens of experience. Each of these traditions offered Kuhn important insights with which to challenge logical positivism.

Interlocutors

Among the earliest letters that Kuhn received after *Structure* was published came from yet another Harvard psychologist, Edwin G. Boring. A past president of the American Psychological Association, Boring had long been a champion of experimental psychology. In 1934, he had been instrumental in convincing Harvard's president, James Conant, to create a department of psychology separate from philosophy, and he had spent much of the 1930s focused on experimental studies of perception. He was among the first researchers in the United States to focus on gestalt images, such as the famous cartoon by W. E. Hill entitled "My Wife and My Mother-in-Law," on which Boring published an article in the *American Journal of Psychology* in 1930. (Boring called such images "ambiguous images" rather than gestalt images, and after his 1930 article they were sometimes referred to as "Boring images" in the United States.) Boring had also written extensively on the history of the field, including his landmark *History of Experimental Psychology*, first published in 1929 and expanded in 1950.[31]

Boring had retired in 1949, though he remained active on campus until his death in 1968. Kuhn and his wife Kay had known Boring's assistant during Kuhn's graduate student days, and Kuhn sent a copy of *Structure* to Boring on its release. Boring was immediately impressed, and he dashed off a letter to Kuhn in early November 1962. "Your book is brilliant and deserves to have an enormous effect," Boring wrote. "You say very much that seems to me to be exceptionally wise, and you say it with clarity and charm. Whatever little bit I can do to get people to read you shall be done."[32]

Boring was as good as his word. In a flurry of letters over the next two months, Boring reported that he was revising his upcoming presidential address to the International Congress of Psychology to incorporate Kuhn's notion of paradigms and the abrupt "all-or-none" transitions between them. He also wrote an unsolicited review of *Structure* for the journal *Contemporary Psychology*—a journal he had founded in 1956 but no longer edited—and sent an unsolicited endorsement to the publicity department of the University of Chicago Press, encouraging them to use it in marketing the book. As Boring wrote to Kuhn in late November 1962, "It seems to me probable that the publisher [of *Structure*] does not know that this is a book about psychology. I know that it is, and I can make it clear to psychologists that it is." Kuhn replied immediately to thank Boring for being so supportive, not least because "your interest and support will certainly gain attention for my book among a group that I had hoped, but not altogether expected, to reach. At this moment, there is nothing that interests me more."[33]

Other readers shared Boring's intuition that *Structure* was "a book about psychology." Of the ninety-one correspondents who wrote to Kuhn about the book during the first five years after its publication, the largest share—nearly 20 percent—came from psychology. Even after members of other academic fields began to engage more substantively with the book, including philosophers and sociologists, psychologists still dominated Kuhn's correspondence about the book, comprising nearly twice as large a portion as any other field[34] (see table 4.1).

Kuhn's friend James Jenkins, the psycholinguist who had spent the 1958–59 academic year with him at Stanford's Center for Advanced Study in the Behavioral Sciences, confirmed the trend. "I enjoyed your book immensely and intensely," Jenkins wrote to Kuhn in the spring of 1965. "I stayed up all night reading it and even went out and *bought* a copy!" He was eager to let Kuhn know that other colleagues in his field had also noticed the book. "I don't know whether you know how much your book is being cited and is influencing current thought," Jenkins continued, and he enclosed a recent review from the *Journal of Linguistics* that had quoted extensively from Kuhn's *Structure* to herald a contemporary "paradigm shift" within the field.[35]

Several other psychologists wrote to express similar sentiments. Many proclaimed that reading the book changed how they thought about their discipline, which they argued was in the midst of its own "revolution." They read *Structure* as a template for what it is to be scientific and argued on that basis that psychology had at last become a legitimate science. A few

Table 4.1. Proportion of Kuhn's correspondents from various academic fields.

Field	Correspondents, %	
	1962–1967 (N = 91)	1962–1981 (N = 170)
Psychology	18.7	14.7
History of science	12.1	7.6
Physics	7.7	6.5
History	6.6	4.1
Sociology	5.5	8.8
Economics	5.5	4.7
Biology	4.4	2.9
Theology	3.3	2.9
Philosophy	3.3	5.3
Philosophy of science	3.3	5.9
Medicine	3.3	3.5
Chemistry	3.3	2.4
English	1.1	2.4
Linguistics	1.1	1.2
Art history	1.1	0.6
Mathematics	1.1	1.8
Political science	0	2.9

told Kuhn that they were requiring their students to read the book, and more than one assigned students the task of writing a paper on an episode in the recent history of psychology using the framework articulated in *Structure*.[36] Influential psychologists like Donald Campbell and Abraham Maslow praised the book; as early as the winter of 1962–63, Campbell wrote that *Structure* had become "my favorite way of introducing students to a proper perspective on science," while Maslow reported that his iconoclastic book, *The Psychology of Science* (1966), was "much influenced by your monograph—which I read with fascination and admiration and gratitude."[37] In 1967, Kuhn was invited to deliver an address, "Logic and Psychology," at the annual meeting of the American Psychological Association; there, younger psychologists took the opportunity to tell Kuhn directly how much his book had influenced their thinking. Two years after Kuhn's address, the president of the association featured Kuhn's book in his presidential address.[38]

Kuhn's psychologist-interlocutors did more than just praise the book. Many pressed Kuhn on specific points. Some wondered why he had not made even more explicit use of Piaget's work or terminology, since (to them) the projects were obviously so similar. As one psychologist explained to Kuhn, "You use assimilation to cover Piaget's assimilation and accommodation, but you probably do not feel this is a problem. According to Piaget, acquiring mastery of a new exemplar would be predominantly accommodation. Solving new problems by means of it would be primarily assimilation. Most problems are a bit of both, which is one way things evolve."[39]

Like the Piaget enthusiasts, several psychologists read Kuhn's book as a treatise on how individual scientists think. Their focus on individual cognition steered their questions. Edwin Boring wondered whether Kuhn thought a paradigm resided in an individual's conscious or unconscious thinking, while a psychologist from Bennington College explained that in his understanding, paradigms referred to an individual researcher's "implicit assumptions, methodological biases and predispositions which lead to ways of thinking and operating." A psychiatrist wrote that Kuhn's *Structure* captured something essential about what it was like to work with psychiatric patients, since the therapist must "work hard with them to change the paradigms in which they construe their experience. In work with a satisfactory patient, one finds repeatedly that there is a succession of revolutions, each of which culminates in a new 'insight'"—an example at the individual level of what Kuhn suggested must occur across an entire scientific community. Like Boring, who was content to gesture that an individual's apprehension of a prevailing paradigm is "in large measure carried in the stream of the Zeitgeist," the psychiatrist saw no need for intervening levels of explanation—no social, institutional, or pedagogical mechanisms that might account for the synecdoche of individual and group.[40]

Kuhn, too, continued to emphasize perception by individual knowers rather than social structures or pedagogical institutions, when pressed by psychologists to clarify what he meant by "paradigm." Boring had gently chided Kuhn that it was "hard for me to pin you down [on the meaning of 'paradigm'] by going through your book," and other psychologists had expressed similar frustrations, all before linguist Margaret Masterman identified twenty-one distinct meanings of the word throughout *Structure*. To Robert Watson, a psychologist at Northwestern University, Kuhn conceded late in 1964 that "I have let myself in for a good deal of misunderstanding by using the term 'paradigm' in too many different ways."[41]

Kuhn found an opportunity to try to clarify what he meant by "paradigm" as he prepared his "Logic and Psychology" lecture for the American Psychological Association meeting in 1967. He sought to emphasize the importance of paradigms as exemplars, concrete problems or techniques that science students learn to manipulate in the course of their training. Kuhn argued that science students must learn to perceive similarities between a new problem and one that they had already learned to solve. Such "a perception of similarity" is "both logically and psychologically prior" to any explicit criteria or definitions that might have been adduced. With an audience of psychologists in mind, Kuhn invoked the distinction between "stimuli" and "sensations," and the "vast amount of neural processing" within an individual's head required to convert the former to the latter. Though such learning processes were irreducibly "neural," Kuhn suggested, they were not entirely "innate": "To an extent still unknown, the production of data from stimuli is a learned procedure."[42]

Rather than illustrate the point with examples from the history of science, Kuhn asked his audience at the American Psychological Association meeting to imagine a child and parent walking around a zoo, as the child learned to distinguish ducks, geese, and swans—not by memorizing definitions or applying explicit criteria, but by repeated exposure to specific examples. The result: "During the afternoon, part of the neural mechanism by which he [the child] processes visual stimuli has been reprogrammed, and the data he receives from stimuli which would all earlier have evoked 'bird' have changed." Here, Kuhn believed, was the essence of how paradigms-as-exemplars worked in science. Several years later, Kuhn's extended example of ducks, geese, and swans from his lecture appeared as the last section of his essay, "Second Thoughts on Paradigms," to illustrate his notion of exemplars.[43]

Kuhn's reliance on categories like perception, stimuli, and sensations clearly resonated with his psychology readers. "Kuhn is wise in choosing perception for his paradigm of paradigms," Boring announced in his essay review for *Contemporary Psychology*, and privately he wondered why Kuhn had not included any gestalt images in the book.[44] Yet others were less convinced. A graduate student in philosophy of science noted that Kuhn's *Structure* "relies heavily on psychology—Gestalt switches, etc.," and wondered whether the perception studies were invoked merely as analogy or as central to his argument. Kuhn replied that his use of gestalt experiments was meant to be metaphorical, "because some of the disanalogies are very severe, especially one's ability to switch back and forth between the two modes of seeing which the diagram allows." But Kuhn averred that the psychological perception studies offered more than mere metaphor: "On the other hand, I do not think that it is merely an accident that fundamental experiments about perception should also illuminate the nature of scientific knowledge and its changes. Doubtless, some of the same fundamental neural processes are involved in both." Unlike the audience at the American Psychological Association, the retreat to individual "neural processes" did not impress the philosophy student, who replied with cautions about overhasty reductionism in the light of the perennial mind-body problem.[45]

Readers from other academic backgrounds responded to different themes within Kuhn's book. He heard from several physicists, for example, some of whom he had known from his earlier studies but most of whom wrote to him, just as most psychologists did, having only encountered his book. None of the physicists asked Kuhn about individual cognition, whether in Piagetian terms or in the framework of neural processing. Instead, several pressed him to elaborate or clarify some of his more sociological suggestions. Mendel Sachs, theoretical physicist at SUNY Buffalo, for example, thought Kuhn's book was primarily about "bandwagons" or fashions in science—that is, the tendency (unfortunate, in Sachs's view) for whole communities of researchers to follow a dominant research trend rather than explore unpopular alternatives. Another physicist wondered if Kuhn was correct that few broadly shared value judgments within a given scientific community persist across scientific revolutions: didn't the community of physicists agree on some measures of what counted as worthwhile research even as particulars of their research programs came and went? Still another physicist congratulated Kuhn for having highlighted so clearly the roles of competing "schools" within a developing scientific field. Of the physicists who wrote to Kuhn, in other words, most read *Structure* as a sociological treatment—and either applauded it or criticized it as such—but none pursued it as an exercise in psychology.[46]

Still, the letters from psychologists kept arriving. Soon after Kuhn's "Logic and Psychology" address he heard from Alfred Fuchs, chair of the Psychology Department at Bowdoin College. Fuchs noted that Kuhn had mentioned in his talk that he had at least two meanings of the word "paradigm" in mind, and Fuchs wondered whether Kuhn had any materials he could share on the topic. Kuhn did not—"I have as yet prepared nothing for publication on this subject"—and he proceeded to sketch for Fuchs the additional clusters of meanings he now wished to clarify:

> One use of the term "paradigm" in my book is to denote the entire matrix of belief and technique shared by the members of a given professional community. I should now prefer to refer to this as something like "the professional matrix," though the phrase does not please me very much. Elements in that matrix would now include such things as: accepted laws, definitions, and other general statements; relevant metaphysical commitments, preferred metaphors, and interpretations of abstract theories; and, as a third major element, paradigms. That use of "paradigms" is the second one which emerges in the book, and, for me, it is the more important one. In this sense, paradigms are restricted entirely to particular concrete examples of the way in

which members of the scientific community in question practice their trade. They are the vehicles of the apprenticeship and practice-problem-solving components of scientific education. In my talk at the A.P.A. they were replaced, for simplicity, by concrete exposure to swans, ducks, and geese.

Over the next year and a half, Kuhn would expand this outline into his famous 1969 postscript. Most of the phrasing stayed the same, though he edited "professional matrix" to "disciplinary matrix."[47]

Kuhn, *Structure*, and the Social

Kuhn's 1969 postscript to *Structure*, with its description of a "disciplinary matrix," moved concertedly beyond a description of individual knowers, let alone neural processing within a given brain. The postscript serves as a useful reminder that Kuhn was not formally trained in psychology, nor were his academic interests limited to that field. Far from it—he read eclectically from his early years, and part of the appeal of *Structure* was surely the fact that the book spoke so eloquently, and concisely, about a range of interesting ideas. Many scholars other than psychologists picked up *Structure* and pursued its ideas, sometimes (as in the case of certain sociologists of science) in directions that drove Kuhn to distraction; certainly not all readers found only lessons about individual cognition in the book.

Nevertheless, Kuhn actively sought out psychologists as an audience, even asking some practitioners (beyond Boring) to review the book in psychology journals to make sure it reached other psychologists' attention.[48] When psychologists did find the book, there was plenty of material to interest them, in an idiom that seemed familiar.

Kuhn's appropriation of insights from specific psychological traditions—developmentalists like Piaget and Werner, "New Look" experimentalists like Bruner and Postman, and psychoanalysis—provided him with impressive material with which to challenge the logical positivists: a neutral observation-language seemed impossible; identical retinal images could produce wildly different concepts and conclusions; preexisting expectations (regarding the colors of suits on playing cards, for example) could affect the "neural processing" within an individual's head, as "stimuli" became "sensations." The Gestaltists' duck-rabbit flips and the way a child learns to sort ducks from geese and swans provided potent resources on which to build an antipositivist, noncumulative account of science.

Late in life, during a lengthy and fascinating interview with three historians and philosophers of science, Kuhn said that his goal as an historian had always been "to get inside a person's head"—always in the singular, and always the head rather than, say, scientists' bench-top apparatus, social groups, or institutional settings.[49] When he defended his position against Popper's in his essay "Logic of Discovery or Psychology of Research," Kuhn wrote that Popper dismissed "the psychological drives of individuals" as irrelevant, yet, Kuhn insisted, a proper understanding of scientific practice "must, in the final analysis, be psychological or sociological. It must, that is, be a description of a value system, an ideology, together with an analysis of the institutions through which that system is transmitted and enforced"—the types of social and institutional trappings that he would later invoke under the banner of "disciplinary matrix." Though he could label them and gesture toward their importance, Kuhn conceded that he would go no further. At that point, he wrote, "my sense that I control my subject matter ends."[50]

Kuhn's road toward *Structure*, and his fruitful discussions with psychologists after the book appeared, thus help to make sense of later historians' somewhat ambivalent relationship with the book. Several historians, like myself, have found inspiration in Kuhn's frequent mentions of the formative role of scientists' training or the centrality of textbooks, only to search in vain throughout his writings for any detailed historical analysis of such critical features of scientific life. Instead, the year after Kuhn's "Second Thoughts on Paradigms" appeared in print, a different model for such studies became available. Michel Foucault's *Discipline and Punish* offered, in its own way, what Kuhn had called for in his critique of Popper: a potent analysis of the institutions through which various "value systems," "ideologies," and practices are "transmitted and enforced." Combining insights from Foucault as well as Kuhn, a new generation of historians of science has sought to make sense of science in this altogether more fleshy idiom: beyond the heads of individual knowers, attuned to the culturally specific, politically situated infrastructures in which knowledge has been produced, mesmerized by the ever-shifting entanglement of ideas and institutions.[51]

Notes

Acknowledgments: I am grateful to Lorraine Daston and Robert Richards for their invitation to participate in the workshop and this volume, and to Lorraine Daston, Michael Gordin, Stefan Helmreich, Erika Milam, Edward Schiappa, and K. Brad Wray for helpful comments on an earlier draft.

1.　Thomas Kuhn, "Reflections on My Critics," in *Criticism and the Growth of Knowledge*, ed. Imre Lakatos and Alan Musgrave (New York: Cambridge University Press, 1970), 231–78 ("I am tempted" quote, 231); Kuhn, "Second Thoughts on Paradigms," in

The Structure of Scientific Theories, ed. Frederick Suppe (Urbana: University of Illinois Press, 1974), 459–82, ("monitoring conversations" quote, 459); and Thomas Kuhn to Charles Christenson, August 17, 1973, Box 4, Folder 8, Thomas S. Kuhn Papers (hereafter, TSK), MC240, MIT Library, Institute Archives and Special Collections ("shudder as I discover" quote). Cf. Roland Barthes, "The death of the author," in Barthes, *Image, Music, Text*, trans. Stephen Heath (New York: Hill and Wang, 1977), 142–148. Barthes's essay originally appeared in 1967. As noted in the preface to *Criticism and the Growth of Knowledge*, the London symposium was held in July 1965 and Kuhn completed his "Reflections" essay in 1969. Likewise, although "Second thoughts" was first published in 1974, Kuhn had completed the text in 1969. See, e.g., Michael Barkun to Thomas Kuhn, 17 September 1969, Box 4, Folder 6, TSK.

2. Kuhn had written the 1969 postscript as an introduction to the Japanese translation of the first edition, but realized it could be helpful for English readers, too: Kuhn, "Postscript—1969," in Kuhn, *The Structure of Scientific Revolutions*, 4th ed. (Chicago: University of Chicago Press, 2012), 173n1. On Kuhn's other attempts to clarify his original intentions, see esp. Kuhn, *Essential Tension*; Kuhn, "Afterwords," *World Changes: Thomas Kuhn and the Nature of Science*, ed. Paul Horwich (Cambridge, MA: MIT Press, 1993), 311–41; and Thomas Kuhn, *The Road since Structure: Philosophical Essays, 1970–1993, with an Autobiographical Interview*, ed. James Conant and John Haugeland (Chicago: University of Chicago Press, 2000).

3. The relevant correspondence may be found in Box 4, Folders 6–16, TSK. These folders contain correspondence specifically in response to *Structure*, organized alphabetically by correspondents' last names. Even in cases in which Kuhn corresponded with particular people about other topics, the correspondence related to *Structure* appears in this location. For example, most of Kuhn's correspondence with his long-time friend, the physicist H. Pierre Noyes, is in Box 11, Folder 51, TSK, except for their letters about *Structure*, which are in Box 4, Folder 13.

4. On Kuhn's use of a dictaphone to compose his letters see, e.g., Thomas Kuhn to Carl Swanson, January 21, 1963, Box 4, Folder 15, TSK; and Kuhn to Dennis A. Rohatyn, January 6, 1969, Box 4, Folder 14, TSK. On dialogic renderings of the meanings of a text, see Mikhail Bakhtin, *The Dialogic Imagination*, ed. Michael Holquist, trans. Caryl Emerson and Michael Holquist (Austin: University of Texas Press, 1981).

5. Edwin G. Boring to Thomas Kuhn, December 7, 1962, Box 4, Folder7, TSK.

6. On sales and translations, see Lawrence Van Gelder, "Thomas Kuhn, 73: Devised Science Paradigm," *New York Times*, June 19, 1996. On journal sections, see *Modern Intellectual History* 9, no. 1 (April 2012): 73–147; *Social Studies of Science* 42, no. 3 (June 2012): 415–80; and *Historical Studies in the Natural Sciences* 42, no. 5 (November 2012): 476–580.

7. On initial print run, see "Book estimate and release" form, dated February 25 [?], 1962, available at the Conceptual and Historical Studies of Science website, http://chss.uchicago.edu/Materials_from_Kuhn_book_files.pdf (accessed February 6, 2013). Kuhn's author questionnaire for the University of Chicago Press, dated March 25, 1962, is available at the same site. Sales figures for 1962–64 from Ian Hacking, "Introductory Essay," in Kuhn, *Structure* , xxxvii. Unless otherwise noted, all quotations from *Structure* will be from the 4th edition.

8. Garrett Hardin to Thomas Kuhn, December 14, 1962, and Kuhn to Hardin, January 21, 1963, both in Box 4, Folder 10, TSK. Kuhn repeated the phrase about his sticking his "neck very far out" in Kuhn to Carl Swanson, January 21, 1963, Box 4, Folder 15, TSK; Kuhn to Donald Campbell, October 22, 1963, Box 4, Folder 8, TSK;

and Kuhn to George J. Stigler, October 24, 1963, Box 4, Folder 15, TSK. Kuhn's correspondence throughout 1963 makes repeated references to the delays in receiving his mail while overseas. The reprint Hardin had enclosed was Garrett Hardin, "The Competitive Exclusion Principle," *Science* 131 (April 29, 1960): 1292–97. See also Hardin, "The Tragedy of the Commons," *Science* 162 (December 13, 1968): 1243–48.

9. Among the faculty who reported to Kuhn on their use of his book in their teaching, see also Francis G. Haber (history department, University of Florida at Gainesville) to Kuhn, December 22, 1964, Box 4, Folder 10, TSK; J. Bruce Brackenridge (physics department, Lawrence University, Wisconsin) to Kuhn, May 23, 1967, Box 4, Folder 7, TSK; Robert A. Lufburrow (physics department, St. Lawrence University) to Kuhn, June 9, 1967, Box 4, Folder 11, TSK; John J. Beer (history department, University of Delaware at Newark) to Kuhn, February 8, 1968, Box 4, Folder 7, TSK; Ronald Abler (geography department, Pennsylvania State University) to Kuhn, April 10, 1968, Box 4, Folder 6, TSK; Thomas C. Cadwallader (psychology department, Indiana State University) to Kuhn, August 19, 1970, Box 4, Folder 8, TSK; James G. Blight (psychology department, Grand Valley State College, Michigan) to Kuhn, February 5, 1979, Box 4, Folder 7, TSK; George Goodin (English department, Southern Illinois University at Carbondale) to Kuhn, November 16, 1979, Box 4, Folder 9, TSK. Abler and Blight mentioned in their letters that they had sent their students' papers to Kuhn.

10. Peter Fox to Thomas Kuhn, January 9 [no year printed, mid-1960s], Box 4, Folder 9, TSK. On annual sales, see Dorothy B. Ebersole (Kuhn's assistant) to James J. Jenkins, September 11, 1975, Box 4, Folder 10, TSK. On blurred boundaries at the time between textbooks, scholarly monographs, and popular books, see David Kaiser, "A Tale of Two Textbooks: Experiments in Genre," *Isis* 103 (2012): 126–38.

11. Diania Jackson to Mack Printing Company (Easton, Pennsylvania, publisher of *American Scientist*), October 14, 1969 ("Kuhn's book is being used" quote), included as enclosure in Derek J. de Solla Price to Diania Jackson (Fairhaven College), November 3, 1969 ("horrifed" quote), copies of which are in Box 4, Folder 13, TSK.

12. Thomas Kuhn to J. Bruce Brackenridge, May 26, 1967, Box 4, Folder 7, TSK ("increasingly and painfully realized" quote); Kuhn to Marvin Barsky, March 2, 1970, Box 4, Folder 6, TSK ("getting perhaps a letter a week" quote); Kuhn to Terry R. Kandall, June 20, 1973, Box 4, Folder 8, TSK. Kuhn sent a letter with identical wording to George Dalton on August 16, 1973, Box 4, Folder 8, TSK. On citations, see Eugene Garfield, "A Different Sort of Great-Books List: The 50 Twentieth-Century Works Most Cited in the *Arts & Humanities Citation Index*, 1976–1983," *Current Contents* 16 (April 20, 1987): 3–7.

13. John H. Van Vleck to Saul G. Cohen, July 2, 1953, in Box 3, Folder "1953–54 Recommendations (A-L)," collection number UAV 691.10, Department of Physics collection, Harvard University Archives, Cambridge, MA. My brief account of Kuhn's composition of *Structure* owes much to Joel Isaac, *Working Knowledge: Making the Human Sciences from Parsons to Kuhn* (Cambridge, MA: Harvard University Press, 2012), ch. 6, as well as to Aristides Baltas, Kostas Gavroglu, and Vassiliki Kindi, "A Discussion with Thomas S. Kuhn," repr. in Kuhn, *The Road since Structure*, 255–323.

14. Thomas Kuhn, "What Are Scientific Revolutions?" repr. in Kuhn, *The Road since Structure*, 13–32 ("I was sitting," "some part of the flux" quotes, 16–17). Kuhn's essay was originally published in *The Probabilistic Revolution*, ed. Lorenz Kruger, Lorraine Daston, and Michael Heidelberger (Cambridge, MA: MIT Press, 1987), 1:7–

20. Kuhn highlighted the "Aristotle experience" in Baltas, Gavroglu, and Kindi, "A Discussion with Thomas S. Kuhn," 275–76. See also Kuhn, *Structure*, xxxix; and George Reisch's, chapter in this volume, "Aristotle in the Cold War: On the Origins of Thomas Kuhn's *Structure of Scientific Revolutions*."

15. Isaac, *Working Knowledge*, 215–17.

16. Jean Piaget, *The Child's Conception of Physical Causality*, trans. Marjorie Gagain (London: Routledge & Kegan Paul, 1930), 24 ("air pushes it"), 238 ("If the child's mind"); see also 20–23, 117–18, 237–40. On Piaget's work and influence, see Fernando Vidal, *Piaget before Piaget* (Cambridge, MA: Harvard University Press, 1994).

17. Jean Piaget, *The Child's Conception of Movement and Speed*, trans. G. E. T. Holloway and M. J. MacKenzie (1946; New York: Basic Books, 1970), 29–30, 32; Kuhn, *Structure*, 118.

18. Herman Witkin, "Heinz Werner, 1890–1964," *Child Development* 36 (1965): 306–28, on 309–10; Mitchell Ash, *Gestalt Psychology in German Culture, 1890–1967: Holism and the Quest for Objectivity* (New York: Cambridge University Press, 1995), 317; Gordon Allport, foreword to Heinz Werner, *Comparative Psychology of Mental Development*, 2nd ed. (New York: Follett, 1948), ix–xii. The first German edition of *Einführung in die Entwicklungspsychologie* appeared in 1926. On Allport and the founding of the Department of Social Relations, see Issac, *Working Knowledge*, 175–76, 180–81.

19. Werner, Comparative Psychology of Mental Development, 22; see also 5, 15–17.

20. Kuhn, *Structure*, xl–xli; Kay Kuhn, personal communication at the Boston University symposium on the anniversary of *Structure*, March 2012; Thomas Kuhn to Richard Gordon, May 4, 1965, Box 4, Folder 9, TSK.

21. Jerome Bruner and Leo Postman, "On the Perception of Incongruity: A Paradigm," *Journal of Personality* 18 (1949): 206–23; Kuhn, *Structure*, 62–64. See also Jerome Bruner, *In Search of Mind: Essays in Autobiography* (New York: Harper and Row, 1983), 67–71, 85–86. On Bruner and the emergence of "cognitive psychology" at the time, see also Jamie Cohen-Cole, "Instituting the Science of Mind: Intellectual Economies and Disciplinary Exchange at Harvard's Center for Cognitive Studies," *British Journal for the History of Science* 40 (2007): 567–97; Cohen-Cole, "The Creative American: Cold War Salons, Social Science, and the Cure for Modern Society," *Isis* 100 (2009): 219–22; and Cohen-Cole, *The Open Mind: Cold War Politics and the Sciences of Human Nature* (Chicago: University of Chicago Press, 2014), ch. 6–7.

22. Isaac, *Working Knowledge*, 219–21. On Frank and the "Unity of Science" movement, see also Gerald Holton, "From the Vienna Circle to Harvard Square: The Americanization of a European World Conception," in *Scientific Philosophy: Origins and Developments*, ed. Friedrich Stadler (Boston: Kluwer, 1993), 47–73; and Peter Galison, "The Americanization of Unity," *Daedalus* 127 (1998): 45–71.

23. Kuhn, *Structure*, 64n13; Baltas, Gavroglu, and Kindi, "A Discussion with Thomas S. Kuhn," 296–98, 300; and K. Brad Wray, "Kuhn and the Discovery of Paradigms," *Philosophy of the Social Sciences* 41 (2011): 380–97 (quote on 384).

24. James Jenkins to Thomas Kuhn, n.d. (ca. April 1965) ("good old days"), and Jenkins to Kuhn, September 9, 1975, Box 4, Folder 10, TSK. The early draft of *Structure*, dated 1958–60, may be found in Box 4, Folder 5, TSK.

25. Thomas Kuhn to Harley Shands, March 12, 1963, Box 4, Folder 15, TSK

26. Baltas, Gavroglu, and Kindi, "A Discussion with Thomas S. Kuhn," 280 ("sure as hell is interesting"); Thomas Kuhn to Maxwell Gitelson, December 18, 1964, Box 4, Folder 9, TSK ("psychoanalysis as therapy"). On Kuhn's early experiences with psy-

choanalysis, see also Jensine Andresen, "Crisis and Kuhn," *Isis* 90 (1999): S43–67; and John Forrester, "On Kuhn's Case: Psychoanalysis and the Paradigm," *Critical Inquiry* 33 (2007): 782–819, esp. 784–89. My thanks to Brad Wray for bringing Forrester's paper to my attention.

27. Kuhn, *Structure*, 64.

28. Ibid., 85–86.

29. Ibid., 126.

30. Ibid., 128.

31. Edwin G. Boring, "A New Ambiguous Figure," American Journal of Psychology 42 (1930): 444–45; Boring, A History of Experimental Psychology, 2nd ed. (New York: Appleton, 1950); Boring, Sensation and Perception in the History of Experimental Psychology (New York: Appleton, 1942); S. S. Stevens, "Edwin Garrigues Boring," Biographical Memoirs of the National Academy of Sciences (1973): 41–76; and Isaac, Working Knowledge, 100–102.

32. Edwin G. Boring to Thomas Kuhn, November 9, 1962, Box 4, Folder 7, TSK. On the Kuhns' relationship with Boring's secretary, Edith Annin, see Kuhn to Boring, November 29, 1962, and Annin's postscript added to Boring to Kuhn, December 7, 1962, Box 4, Folder 7, TSK.

33. Boring to Kuhn, November 26, 1962 (including a draft of Boring's upcoming presidential address); Kuhn to Boring, November 29, 1962, Box 4, Folder 7, TSK. Boring sent his endorsement to the publicity department of the University of Chicago Press (with a copy to Kuhn) on November 26, 1962, and sent a draft of his review for *Contemporary Psychology* to Kuhn on December 4, 1962. In all, Boring and Kuhn exchanged ten letters between November 9 and December 18, 1962, all in Box 4, Folder 7, TSK.

34. *Structure* was published in September 1962, and correspondence in response to the book began several weeks later; hence, the period labeled "1962–67" includes five years and two months. Most correspondents wrote to Kuhn on letterhead stationery, which clearly identified their institutional and departmental affiliations. For all but a handful of the remaining letters, unambiguous identifications could be made from the contents of the letters or titles listed below signatures. No identifications were possible for a modest fraction of correspondents (2.2% between 1962 and 1967, 6.5% between 1962 and 1981). In addition to the fields listed in table 1, other groups include students (4.4% during 1962–67, 5.9% during 1962–81); university administrators (3.3% during 1962–67, 2.4% during 1962–81); and lawyers (1.1% during 1962–67, 1.2% during 1962–81). In table 1, I have distinguished historians of science from general historians. During the early period (1962–67), many of the letters Kuhn received from historians of science were brief acknowledgments from close colleagues of receipt of the book rather than substantive letters, and hence quite different in character from the letters Kuhn received from general historians. See, e.g., Harry Woolf to Thomas Kuhn, December 11, 1962, Box 4, Folder 16, TSK; Charles Gillispie to Thomas Kuhn, January 21, 1963, Box 4, Folder 9, TSK; Alexandre Koyré to Thomas Kuhn, August 9, 1963, Box 4, Folder 11, TSK; Melvin Kranzberg to Thomas Kuhn, July 19, 1965, Box 4, Folder 11, TSK; Derek J. de Solla Price to Thomas Kuhn, December 16, 1965, Box 4, Folder 13, TSK.

35. Jenkins to Kuhn, n.d. (ca. April 1965). Jenkins enclosed a photostat of James P. Thorne's review of Paul Postal's *Constituent Structure: A Study of Contemporary Models of Syntactic Description*, which had appeared in *Journal of Linguistics* 1 (April 1965): 73–76.

36. Kenneth Hammond to Thomas Kuhn, February 15, 1963, Box 4, Folder 10, TSK; David Smith to Kuhn, June 15, 1964, Box 4, Folder 15, TSK; Robert Lissitz to Kuhn, September 28 1964, Box 4, Folder 11, TSK; Stewart Perry to Kuhn, December 2, 1964, Box 4, Folder 13, TSK; Alfred Fuchs to Kuhn, September 15, 1967, Box 4, Folder 9, TSK; Robert Ravich to Kuhn, September 17, 1967, Box 4, Folder 14, TSK; Danny Moates to Kuhn, January 2, 1969, Box 4, Folder 12, TSK; Wayne Lazar to Kuhn, March 7, 1969, Box 4, Folder 11, TSK; Walter Weimer to Kuhn, October 30, 1969, Box 4, Folder 16, TSK; Jacobo Valero to Kuhn, March 17, 1970, Box 4, Folder 16, TSK; and Thomas Cadwallader to Kuhn, August 19, 1970, Box 4, Folder 8, TSK.

37. Donald Campbell to Thomas Kuhn, n.d. (ca. January 1963), Box 4, Folder 8, TSK; Abraham Maslow to Kuhn, n.d. (ca. January 1967), Box 4, Folder 12, TSK. See Abraham Maslow, *The Psychology of Science: A Reconnaissance* (New York: Harper and Row, 1966).

38. On Kuhn's address at the American Psychological Association, see Fuchs to Kuhn, September 15, 1967; and Cadwallader to Kuhn, August 19, 1970. See also George Miller, "Psychology as a Means of Promoting Human Welfare," *American Psychologist* 24 (1969): 1063–75, esp. 1066, 1070, which reproduces Miller's September 1969 presidential address to the American Psychological Association.

39. Jane Loevinger to Thomas Kuhn, April 21, 1969, Box 4, Folder 11, TSK; see also Harley Shands to Kuhn, January 30, 1963, Box 4, Folder 15, TSK.

40. Boring to Kuhn, December 7, 1962; Louis Carini to Kuhn, June 30, 1963, Box 4, Folder 8, TSK; Harley Shands to Kuhn, March 15, 1963, Box 4, Folder 15, TSK.

41. Boring to Kuhn, December 7, 1962; Thomas Kuhn to Robert Watson, October 27, 1964, Box 4, Folder 16, TSK. See also Carini to Kuhn, June 30, 1963. Cf. Margaret Masterman, "The Nature of a Paradigm," in Lakatos and Musgrave, *Criticism and the Growth of Knowledge*, 59–89. Masterman first delivered her paper in July 1965 at the International Colloquium in the Philosophy of Science in London.

42. Quotations from Kuhn, "Second Thoughts on Paradigms," 308–9. It appears that Kuhn's talk at the 1967 American Psychological Association meeting is no longer extant in his papers, but from the descriptions of the talk in correspondence from the time, it seems clear that much of that presentation was included in his essay "Second Thoughts," the first complete draft of which dates from ca. 1969, even though it was not published until 1974. For discussion of the APA lecture, see Fuchs to Kuhn, September 15, 1967, and Kuhn to Fuchs, September 25, 1967, Box 4, Folder 9, TSK. On references to drafts of "Second Thoughts," see also Jane Loevinger to Thomas Kuhn, April 21, 1969, Box 4, Folder 11, TSK; Michael Barkun to Thomas Kuhn, September 17m 1969, Box 4, Folder 6, TSK; Kuhn to Bruce Kuklick, March 2, 1970, Box 4, Folder 11, TSK. Kuhn cited the essay as "in press" in Kuhn, "Postscript—1969," 174n3. On the evolution of Kuhn's meanings of "paradigm," see also Wray, "Kuhn and the Discovery of Paradigms."

43. Kuhn, "Second Thoughts on Paradigms," 309–10.

44. Boring, draft essay review of *Structure*, enclosed in Boring to Kuhn, December 4, 1962 ("Kuhn is wise"); Boring to Kuhn, November 26, 1962.

45. George Ballester to Thomas Kuhn, May 26, 1972 ("relies heavily on psychology"); Kuhn to Ballester, June 9, 1972 ("disanalogies are very severe"; "On the other hand"); Ballester to Kuhn, July 7, 1972, Box 4, Folder 6, TSK.

46. Mendel Sachs to Thomas Kuhn, September 10, 1970, Box 4, Folder 15, TSK; T. E. Phipps to Kuhn, June 25, 1963, Box 4, Folder 13, TSK; G. Jona-Lasinio to Kuhn,

July 23, 1973, Box 4, Folder 10, TSK. See also Leonard Loeb to Kuhn, February 27, 1963 and April 15, 1963, Box 4, Folder 11, TSK.

47. Kuhn to Fuchs, September 25, 1967; Kuhn, "Postscript—1969." One of the best explications of Kuhn's two main clusters of meanings of "paradigm" remains Joseph Rouse, "Science as Practice: Two Readings of Thomas Kuhn," in Rouse, *Knowledge and Power: Toward a Political Philosophy of Science* (Ithaca, NY: Cornell University Press, 1987), 26–40.

48. See, e.g., Kuhn to Shands, March 12, 1963.

49. Baltas, Gavroglu, and Kindi, "A Discussion with Thomas S. Kuhn," 276, 280.

50. Thomas Kuhn, "Logic of Discovery or Psychology of Research?" in Lakatos and Musgrave, *Criticism and the Growth of Knowledge*, 1–23 (quote, 21). See also Forrester, "On Kuhn's Case," 789–90, 798–99.

51. Michel Foucault, *Discipline and Punish*, trans. Alan Sheridan (1975; New York: Pantheon, 1977). See also, e.g., Andrew Warwick and David Kaiser, "Kuhn, Foucault, and the Power of Pedagogy," in *Pedagogy and the Practice of Science: Historical and Contemporary Perspectives*, ed. David Kaiser (Cambridge, MA: MIT Press, 2005), 393–409; Cyrus Mody and David Kaiser, "Scientific Training and the Creation of Scientific Knowledge," in *Handbook of Science and Technology Studies*, rev ed. (Cambridge, MA: MIT Press, 2007), 377–402.

FIVE

Paradigms

IAN HACKING

My introductory essay for the fiftieth anniversary edition of *Structure* de-
votes more words to paradigms than to any other topic.[1] One justification
is what Kuhn himself wrote in his "Postscript —1969": "The paradigm as
shared example is the central element of what I now take to be the most
novel and least understood aspect of this book."[2] Many careful readers
agree and have been much helped by his subsequent elucidations, but continue to find the paradigm idea perplexing. This chapter urges that perplex-
ity is in the nature of the beast, and moreover, that the difficulties go back
to ancient times and have never been fully resolved.

In a conference paper written about the same time as his postscript and
titled "Second Thoughts on Paradigms," Kuhn said that he had used the
word "paradigm" in two main ways, one local and one global. In the "Postscript," the first three sections are titled "(1) Paradigms and community
structure," "(2) Paradigms as the constellation of group commitments,"
also spoken of as a 'disciplinary matrix,' and (3) The "central element" of
novelty in the book, "Paradigms as shared examples." These are also described as one type of element in the disciplinary matrix.

The constellation of group commitments is global—and is Norton
Wise's focus in his chapter in this volume. The shared examples, on the
other hand, are local, and form Lorraine Daston's focus. I shall be concerned only with (3), paradigms as shared examples (or, as Kuhn sometimes preferred, exemplars). The discussion will be embedded within more
general reflections on the history of reasoning.

What follows is philosophy, philosophical logic, perhaps, but definitely
not philosophy of science, let alone history and philosophy of science. It
is, however, a glimpse at the history of (mostly Western) thinking about

arguments—starting, as I did in the "Introductory Essay," with Aristotle's *paradeigma* in the *Rhetoric*.

When There Are No Rules or Systems to Guide Us

Grammatical paradigms—*amo, amas, amat*—in which a verb is conjugated or a noun declined as a pattern to be followed for any root of the same form are relatively unproblematic, although one has to learn the exceptions. The pupil does see how to go on, and the formal statement of a rule may accompany the paradigm. Philosophers have gnawed away at problems of following rules that troubled Ludwig Wittgenstein and prompted Saul Kripke to rock quite a few boats; but in real life schoolchildren of old, who had to master the paradigms, had no difficulty in doing so except, in many cases, encountering boredom amounting to revulsion. The exemplar of the conjugation "to love" was far easier to mimic than to follow an explicit rule conveying the same information.

Otherwise, paradigms, examples, exemplars, analogies, resemblances, models, and their ilk present a big problem for those of us who are schooled in deductive logic. They don't much work by rules. This is elegantly captured in the title of a valuable paper by a professor of rhetoric, John Arthos: "Where there are no rules or systems to guide us: argument from example." This is not about what analytic philosophers call "the rule following considerations," which can be cast in the form of a regress argument to the effect that the giving of rules has to come to an end—and not with a rule. What Arthos points to is that the use of exemplars in argument does not appear to be rule governed at all. It is not that the giving of rules had to come to an end; it does not start. You have to see the point of the example, or learn how to exploit a model, or draw an analogy, case by case.[3]

Kuhn's Three Self-Explications

On more familiar terrain for readers of *Structure*, Margaret Masterman's paper, the one that delineated twenty-one different ways in which Kuhn used the word "paradigm," concluded that what Kuhn owed us is a new explication of "argument by analogy."[4] And Kuhn *did* do that in his "Second Thoughts," and in three different ways.

(A) One of them is a sort of metaexemplar, a lovely example of how an exemplar can be exploited. It starts with Galileo's work on the inclined plane and moves on to Huygens's studies of the pendulum and finally to

Bernoulli's discovery of "how to make the flow of water from an orifice in a storage tank resemble Huygens' pendulum." This kind of use of exemplars seems to require nothing short of genius. Of course it is a long series of try-outs and explorations by various groups of people, some plodding, some nimble. This particular paradigm initiated by Galileo applies to a group of cases, each drawing an analogy with other, usually earlier, members of the group. Stanislav Smirnov won a 2010 Fields Medal for work in percolation theory, which can be seen as a continuation of analogies going as far back as Galileo. But another medal awarded that same year, to Ngô Bao Châu for proving the fundamental lemma of the Langlands program, derives from a quite different paradigm. On the other hand, the story from Galileo's rolling balls to Bernoulli is a metaexample of the use of analogy and modeling in the sciences, a completely general phenomenon.

Mary Hesse's *Models and Analogies in Science* (1963), published directly after *Structure* and almost uninfluenced by it, is a survey of different ways in which analogies work in the sciences. Such work has continued, for example, by Rom Harré.[5] A recent important contribution is Paul Bartha's *By Parallel Reasoning* (2010), subtitled *The Construction and Evaluation of Analogical Arguments*. This is, incidentally, a valuable critical resource for philosophers curious about various computer modelings of analogical arguments.

(B) Kuhn's reflections on the role of the problems at the end of the chapters in a textbook were a quite different approach to clarifying the idea of a paradigm. That is where the student learns how to do physics. It becomes mere habit to apply the same method of solution to one problem, which worked in a similar problem. Here the exemplar is most like a paradigm in grammar. Indeed, the weaker students learn only a rote way to solve some problems, without being well able to branch out on new but related issues. There may even be a few rules of thumb to guide the more challenged students, although the able ones will at once see how to go on. That is how ability is recognized and encouraged during the larger process of initiation. The good students catch on quickly and can advance up the ladder.

The importance of problems at the end of textbook chapters was a really important insight. But one should be cautious about generalizing. That is how physics and some other sciences and technologies were in fact taught and learned in those days. (That's how *I* learned physics as an undergraduate.) Textbook writing and use has changed, and what Kuhn saw was only a temporal slice of an ongoing evolution of curricula. Those incredibly expensive, heavy, illustrated tomes, making a fortune for the authors and publishers of central courses, are very distinctly on the way out, and electronic devices, hypertext, and its successors will become more prevalent, especially as global online education takes hold. But Kuhn's insight does have staying power, for it is still the case that solving problems by pencil and paper is an essential part of physics apprenticeship, and also engineering and other similar parts of technoscience.

(C) A final and very different attempt to clarify the notion of a paradigm is to be found in Kuhn's "Second Thoughts." This is once again a metaexample, the example of Johnny learning how to use the words "swan," "duck," and "goose" from examples. It is a lesson in ostensive definition, by giving examples and counterexamples, by means of which the father teaches Johnny the use of some words. It assumes that Johnny already recognizes what kind of thing (birds, in this case, or, at any rate, living creatures) is being named. Kuhn seems to have thought that this metaexample was his most important contribution to clarifying the idea of a paradigm.[6]

The Johnny example was later to be much emphasized by David Bloor and then Barry Barnes and many others in the science studies field, and not just by adherents of the strong programme of the Edinburgh school. They have been called the "left Wittgensteinians." They connect Johnny with the "rule-following considerations."[7] In my own nonstandard opinion, this (ab)use of ostensive learning has no essential connection to the important insights of the strong programme. It is perhaps worth rereading Wittgenstein's own onslaught on a particular theory of naming by ostension with which he begins the *Philosophical Investigations*[8]—and then applying it to the Johnny example.

(A) Successive solutions, Galileo & co., (B) problem sets, and (C) Johnny: they don't seem to have much to do with each other. Kuhn may have thought that (C) enabled us to understand (A) as a simple, nonhistorical instance of the same process. He may also have thought that (C) is an instance of a process used in (B) mastering problem sets. All three work around something that we do not well understand and for which we use those words: models, analogies, resemblances, metaphors . . . and paradigms or exemplars.

The Paradeigmatic Muddle

As I understand Daston's contribution to this volume, she holds that *nobody* has ever succeeded in giving a clear explication of Kuhn's doctrine of the paradigm as exemplar. We might speak of the "paradeigmatic muddle." The "muddle" is the name I give to our persistent inability to get clear about something (some things?) over which words such as "analogy" hover. My label "paradeigmatic" references Aristotle's use of the word *paradeigma*, translated as *exemplum* into Latin, and often as "example" or "exemplar" in English. I don't call it "paradigmatic," for that would commit me to meaning that the muddle was a paradigm of muddles (but maybe it is!).

I believe the muddle arose long ago, in connection with Aristotle's *Rhetoric*, and is illustrated by scholarly debate about what Aristotle meant by "paradigms" and their connection with induction. Debate is merited, for Aristotle himself did not (in my opinion) get clear about these matters. Thus, I argue for a "big picture." The problems, with the phenomena Kuhn emphasizes, have been a problem for Western theories of reasoning ever since Aristotle invented deductive logic (the syllogism), *after* he had developed his theory of rhetoric. If we did not have deductive logic as our gold standard, we would not be troubled.

In Aristotle, argument by example is one of the two (and only two!) modes of argument or kinds of reasoning in *Rhetoric*. Kuhn's paradigms are not exactly arguments. (A), the reasoning that leads from Galileo's inclined place to Bernoulli's hydrodynamics, can be thought of as "one long argument." (B), working problems in the back of the book, is reasoning towards the solution of a series of exercises, and can be redescribed as argument (for the right answer). (C), Johnny's ostensive learning, is not an argument at all, and is of course much closer to picking up a grammatical paradigm for a conjugation, than to an argument. But historically, it is in the discussion of rhetorical argument by examples that the muddle first comes to the fore.

Aristotle and *Paradeigma*

It is widely agreed that Aristotle's first (preserved) lecture courses were probably *Topics* (also called *Dialectics*) and *Rhetoric*. Both are about argument, in the sense of trying to convince someone else—and not in the sense of trying to find out the truth or settle on the best course of action. Here is what Aristotle says he is doing in *Rhetoric*: "The function of rhetoric is not to persuade but to see the available means of persuasion in each case."[9]

Aristotle did not invent the word *paradeigma*, but he made very specific use of it in *Rhetoric*. *Topics* is about dialectic, that is, arguments between two parties who can debate matters back and forth. *Rhetoric* is about argument directed at an audience, typically by an orator, but also in a court of law. Since the argument is spoken and audiences have short attention spans, brevity is a virtue for rhetoric, and less so for dialectic, where one can go back over a thought again and again. One often need

not state what is generally believed, and one's examples should be familiar to the audience. In both dialectic and rhetoric we are concerned with
oral argument.

In both books, he wrote of syllogisms (*sulligismos*), but he did not mean
"the syllogism," such as *Barbara* (the so-called first of the four figures of
the syllogism, for example: all men are mortal; Socrates is a man; therefore
Socrates is mortal). The idea of necessarily truth preserving argument, what
we call logical consequence, came later.[10] The formal theory of the syllogism is expounded in *Prior Analytics* A 1–7. This chronology is important to
my analysis, because I claim that later readers, and even Aristotle to some
extent, devalued the *Rhetoric* in favor of the new gold standard, the syllogism that preserves truth. And then, to foreshadow, argument by example
was pretty much dumped in the early modern period, in part because the
new sciences were groping toward a new idea of inductive reasoning. But
that very word, induction (*epagōgē*) has generated a veritable morass of interpretations of Aristotle.

Aristotle begins *Rhetoric* by saying that dialectic and rhetoric are counterparts. Here is his opening statement:

> Rhetoric is a counterpart [*antistrophos*] to dialectic. For both are concerned
> with such things as are, to a certain extent, within the knowledge of all peo-
> ple and belong to no separately defined science. A result is that all people, in
> some way, share in both; for all, up to a point, try both to test and uphold
> an argument [as in dialectic] and to defend themselves and attack [others,
> as in rhetoric.] [. . .] it is possible to observe why some succeed by habit and
> others accidentally, and all would agree that such observation is the activity
> of an art [*teknē*].[11]

They "belong to no separately defined science" and yet their practice is an
art—which thereby can be taught. Notice how oedipal this statement is. It
totally ignores what Socrates had said against the sophists in *Gorgias*, that
only subjects with a subject matter, such as cookery or carpentry, naviga-
tion or medicine, can be taught. *Rhetoric* may have created quite a stink
among older members of the Academy. Note that logic, be it in the form of
the syllogism or what is still called symbolic logic, is like rhetoric in that it
has no specific subject matter. In this they contrast with mathematics, and
in particular with Euclidean geometry, which has a subject matter, namely
shapes such as circles or conic sections. Aristotle never imagines that the
syllogism and geometry could be filed under the same heading. The very
idea that mathematics is logic arrived only in the nineteenth century—an

idea never accepted by most mathematicians, despite its effect on English-language analytic philosophy (thanks to Gottlob Frege and Bertrand Russell). It is perhaps remarkable that both demonstrative proof in geometry and truth-preserving argument in logic were developed in the same civilization, and almost concurrently.

Aristotle castigates previous manuals of rhetoric as being concerned with merely winning arguments by any means. His own work proceeds differently. As the distinguished scholar of rhetoric George A. Kennedy writes, "Aristotle's major objective is clearly an understanding of the nature, materials, and uses of rhetoric."

Rhetoric is spoken to an audience. Audiences have short attention spans. Long arguments are to be avoided. Agreed common knowledge is always the best starting point, and often if the orator is familiar with the audience, then most of it can be assumed, not stated. Arguments should be brief. Dialectic is a back-and-forth argument, so that points can be recalled in the course of discussion. There is not quite the same need for brevity. But in either case, one wants to use, as much as possible, premises that are agreed and do not need stating. Moreover, examples in rhetoric should be well known to the audience, so that once stated they remain in the mind of the authors. Thus, although I have slipped from Kuhn's "paradigms as shared example" to paradigm as example, Aristotle insisted that examples used in rhetoric should be shared between orator and audience.

The Two—and Only Two—Types of Argument

Aristotle makes the *extraordinary* claim that there are exactly two types of argument used in each of dialectic and rhetoric. Dialectic proceeds either by induction (*epagōgē*), or syllogism (*sullogismos*). Rhetoric proceeds either by *paradeigma* or *enthymēma*. These are counterparts indeed:

> for the *paradeigma* ["example"] is an induction, the *enthymēma* a syllogism.
> I call a rhetorical syllogism an enthymeme, and a rhetorical induction a paradigm.[12]

Sounds simple, does it not? The two most obvious readings of this text are long established, but just *wrong*, and here I appeal to authority.

Traditional logic and today's textbooks teach that an enthymeme is an abbreviated syllogism, with a well-known major or minor premise omitted for brevity. That fits with the need for brevity in spoken argument, but Myles Burnyeat establishes conclusively that Aristotle does not mean a classical syllogism such as *Barbara*, with a premise omitted, but rather that a syllogism in the early texts is an argument by stages from which an inference is drawn.[13] An enthymeme is an argument, possibly from many premises, some stated, some not. That does not concern me here, but it is worth having in mind when we turn to paradigms.

As for paradeigma, the standard reading of the text is that the giving of examples corresponds to induction by simple enumeration. The need for brevity in rhetoric forces us to skip the enumeration and give just one or two examples. Arthos calls this the "inductive formulation." He writes most emphatically, "I can discover no exceptions in argument textbooks to this inductive formulation."[14] He thinks it is totally wrong, just as Burnyeat categorically rejects the standard view of enthymemes as abbreviated syllogisms. I am an outsider to both fields of scholarship, but on reading Aristotle I agree with both authors.

The paradeigmatic muddle begins here. Even Burnyeat contributes to it, because he holds that *Rhetoric* was an essay towards a theory of induction. He agrees with me that no plausible account of inductive reasoning was really in the cards until the seventeenth century, with the emergence of probability. He holds that indifference to induction was a result of Stoic admiration for necessarily truth-preserving argument. Be that as it may, Buryeat thinks of Aristotle's idea of enthymeme as nondeductive argument, from which he moves to the idea that nondeductive is inductive. Like nearly all classical scholars, he writes much about enthymeme but almost nothing about examples. A discussion of *paradeigma* is far more likely to appear in a journal directed to rhetoric than one devoted to classical texts. Indeed, journals have an almost disjoint readership, and in the English-speaking world, departments of rhetoric (or speech) have almost no contact with departments of philosophy. One classicist who has taken example seriously is Geoffrey Lloyd, both in his dissertation and in studies comparing ancient Greece and China.[15]

What Did Aristotle Mean by "Induction"?

Many (I think most) readers take him to mean induction by simple enumeration. Well, not exactly. It is rather that we give some instances, and find a genus under which all instances fall. Not all classes are genera, but I shall not try to state the criteria for being a genus.

For an example of a generalization under a genus, we could use John

Stuart Mill's "Tom, Dick and Harry were mortal, so all men are mortal."
Here we bring "man" under the genus "mortal." Aristotle realizes in *Topics*
that such induction does not exhaust nondeductive argument:

> Try to secure admissions by means of likeness. . . . This argument resembles
> induction, but is not the same thing; for in an induction it is the universal
> whose admission is secured from the particulars, whereas in arguments from
> likeness, what is secured is not universal under which all the like cases fall.[16]

Possibly Aristotle had in mind Mill's argument from particulars to particulars: "Tom, Dick and Harry were mortal, so the Duke of Wellington is mortal."[17] But there, there is a universal, "all men are mortal," waiting in the
wings. Aristotle must have had in mind resemblances precisely where there
was no plausible universal generalization to guide us. But he would have
put it more strongly: where there is no relevant genus under which the particular falls, no inference can be drawn.

Aristotle's Examples of Examples

Arthos and a few other readers, including Geoffrey Lloyd, suggest that Aristotle stated two views that were never disentangled, one in which example pretty much is induction by enumeration, abbreviated, and another in
which example is a major form of argument. It is the latter Aristotle that
Arthos and I find interesting. Here is an example of Aristotle's that fits the
inductive formulation:

> Dionysius is plotting tyranny because he is seeking a bodyguard; for Peisistra-
> tus also, when plotting earlier, sought a bodyguard and after receiving it made
> himself tyrant, and Theagenes did the same in Megara, and others, whom the
> audience knows of, all become examples for Dionysius. . . . All these actions
> fall under the same genus, that one plotting tyranny seeks a guard.[18]

This one is not so clearly an enumeration, although two instances are cited:

> It is necessary to make preparations against the King of Persia and not allow
> Egypt to be subdued. For, in the past Darius did not invade Greece until he
> had taken Egypt, but after taking it, he invaded; and again, Xerxes did not at-
> tack until he took Egypt, but having taken it, he invaded; thus if the present
> Persian king takes Egypt, Greece will be invaded and this result must not be
> allowed.[19]

These are historical examples. They concern matters of policy. So we think of them as at least attempts at generalization from past experience. Aristotle notes that the orator must use shared examples, familiar to the audience.

The Invented and the Fabulous

Aristotle then proceeds to something extraordinary, which cannot be reduced to generalization from established or known events. In book II, chapter 20, he begins by saying that examples can be of two kinds. They can be histories, or they can be "inventions of facts by the speaker." These inventions can be either illustrative parallels, or they can be fables.

Our first instinct, today, is to dismiss the invented and the fabulous. *Our inferences are supposed to rely on the facts!* Aristotelian scholarship has remarkably little to say about these types of paradigms, yet it must seem to the outsider that it is essential to understand how they fit into Aristotle's theory of rhetorical argument.

On examination, his illustrative parallels are not so shocking. We ought not choose public officials by lot, for that would be like choosing a navigator by lot from among a ship's crew. The example is something with which any audience would agree, so the speaker hopes that the audience sees the parallel. Notice that there is little temptation to call this an induction.[20]

What about the fables? Aristotle gives two examples. Phalaris was tyrant of what is now Agrigento in Sicily. The people wanted to give him a bodyguard. The long-lived poet Stesichorus (ca. 640–555 BCE) argued that this was a bad idea. Once there was a horse grazing in a field. Then came a stag to share the pasture. The horse asked a man if he could help get rid of the interloper. Sure, said the man, and hopped up on the horse. Now the horse was far worse off, enslaved by the man.[21]

Then we have none other than Aesop, who lived in Samos off the coast of Asia Minor, once the richest Greek city. A popular leader was deemed corrupt and on trial for his life. Well, a fox crossing a river was washed away and stuck in a hole in the bank. She was then infested by fleas. A hedgehog offered to help by picking off the fleas. The fox declined, saying that the fleas she had acquired were gorged and no longer hungry. "These are already full of me and draw little blood, but if you remove these, other hungry ones will come and drink what blood I have left." Aesop continued: "In your case too, O Samians."[22]

"Fables are suitable in deliberative oratory" for it is often easier to find an example from fable than from history. "They should be made in the same way as comparisons, provided one can see the likenesses, which is rather easy from philosophical studies." Nevertheless, "examples from history are more useful in deliberation: for future events will generally be like those of the past."[23]

He goes on to say that it is preferable to argue by enthymemes, but if we wish to combine the two, the examples should follow the enthymeme. They will then be like witnesses supporting the first argument, and a single witness, if it is a good one, is enough. I have observed that Burnyeat takes Aristotle to be grappling with nondeductive argument in the form of induction. Arthos proposes that Aristotle was also, and perhaps less clearly, grappling with nondeductive argument in the form of example, explicitly leading rhetoric away from enthymeme to argument by *paradeigma*.

What does all this teach us about argument by example? Perhaps, as has been suggested, Aristotle had two types of argument in mind, which he never sorted out. One was in some way inductive, although he never clarified his ideas on induction. The other was the parallel, be it fact or fiction. One might offer here the idea that this is an argument from likeness, as in the passage quoted from *Topics* VIII above. But mere likeness won't do. It has to be a *good* parallel. The audience has to find it *enlightening*. And, of course, what strikes one audience as helpful may totally fail for another audience.

I will return to this below, examining recent uses of particular case arguments. But first we notice that Aristotle has left a great many loose ends. He simply had not figured out about nondeductive argument, which is why scholars cannot agree on what theory he held—for surely the Stagyrite held a definitive theory? I am sure he did not, and no wonder. Indeed, as far as example is concerned, it tended to be pushed aside after he had invented the syllogism. Burnyeat attributes the dismissal of the themes of *Rhetoric* to the Stoics. "If one believes that an adequate philosophy of science must find a place for inductive as well as for deductive logic, one will conclude that, as logicians, Aristotle was a better friend to the sciences than Zeno and Chrysippus."[24] He holds that what halted the study of nondeductive argument in mid-stride was Stoic logic, with its fixation on what we now call logical consequence and validity in deductive logic. Arguments that were necessarily truth-preserving displace all others.

Because Burnyeat is interested in induction, he did not attend to the other wing of *Rhetoric*: argument from example. Induction may have been downplayed until early modern times, but example flourished in the Roman world and the Renaissance. Argument by example persisted until the end of the Renaissance.[25] But as it petered out, there arose what has been called "the

Renaissance crisis of exemplarity."[26] This happened in part because of the new attention to induction, but also because no one had ever provided rules for the use of an example that would satisfy a precise logical mind.

Exemplarity was not quite killed off in early modernity, but it was transformed. When you look, you will find that an amazing amount of Bacon still retains example. His *Essays* could be textbook applications of *Rhetoric*, rich in vivid historical or mythical examples familiar to his readers but not to many people today. And what are his "prerogative instances" if not paradigms or exemplars?

Rigolet begins his essay on the crisis of exemplarity with its death knell, an assertion of Montaigne (1533–92), whose *Essays* can be read for many reasons, including their intense critique of rhetoric. The rambling final essay, "On Experience" has the oft-quoted damnation of example, *"Tout exemple cloche"*: "All things hold together by some similarity or other; *every example limps*, and the connexion that is drawn from experience is always faulty and imperfect."[27]

This is the complaint that Nelson Goodman would have made, and did in fact make in his famous "Seven Strictures on Similarity."[28] If we treat example as a logical relation, then there are always too many analogies to be drawn. Aristotle well knew it is not a logical relation; yes, it works through likeness or similarity, but it must be what I have called enlightening likeness, a comparison that opens up a new way of seeing things—to a specific audience. Because it is "relative," no one has given a satisfying analysis of enlightening likeness—never, in the whole history of reasoning about reason.

Yet we argue by example all the time. It is a key element in argument, institutionalized by the use of precedent in the common law, but also everywhere. I will not make the kind of claim cognitive scientists make, that it is a human universal, but one can find it in most societies.

The Taj Mahal

Let us conclude by turning to a few recent philosophers who have emphasized example. I shall begin with a lovely twentieth-century illustration due to John Wisdom (1904–93). It was first used in a talk about gods for the BBC in 1950.

> Imagine someone trying on a hat. She is studying the reflection in a mirror like a judge considering a case. There's a pause and then a friend says in tones too clear, "My dear, it's the Taj Mahal."[29]

"Instantly" . . . the good lady realizes how absurd she looks. Wisdom is at pains to insist that this is not like informing "someone that it has mice or will cost a fortune. It is more like saying to someone 'Snakes' of snakes in the grass but *not* concealed by the grass but so well camouflaged that one cannot see what's before one's eyes."[30]

Importantly, "it isn't true that the words 'It's the Taj Mahal!' meant 'It is like the Taj Mahal.' This more sober phrase is an inadequate substitute. It is feebler than the original and yet too strong. For the hat isn't like the Taj Mahal, it's much smaller, and the shape is entirely different."[31]

Situation is, then, a great deal: The friend speaks "in tones too clear." She is warning, and the lady who was trying on the hat gets the point. It is an enlightening—indeed, a devastating—comparison. It is, among other things, an implicit argument by example, an argument whose conclusion is the imperative: Don't buy the hat! It is more powerful to my ears than any of Aristotle's examples, historical or fabulous. But it is effective— enlightening—only to those who share a vestige of Wisdom's British middle-class 1950s sensibilities. All the philosophers who have admired Wisdom's example are, even if distantly acquainted with such a milieu. One thinks of Cora Diamond, for example: "This is plainly not a case in which communication has broken down, but the way the woman sees the hat may be radically changed."[32] I expect there are many young Brits, let alone people in the larger English-speaking world, who would simply not get the point. Whether or not an example is seen or experienced as en- lightening is relative to the sensibilities and background of the audience.

Aristotle well knew this: the orator must pick examples that are well- known to the audience. His historical and fabulous examples were well- known in his day, but need some explanation, even to today's well-educated reader.

The cardinal fact is that these examples, these *paradeigma*, must be en- lightening, and that depends on the audience. They are, to use jargon, con- text sensitive. That is not the sort of thing that logic has been good at ana- lysing; the closest we have come is theory of indexicals. For indeed there are "no rules or systems to guide us."

What's So Great about Examples?

I will answer by quoting some of Arthos's own use of authorities.[33] Hans- Georg Gadamer implicitly addressed the error of the "inductive formula- tion," the error of confusing induction with practical arguments using ex- amples, when he wrote:

The individual case does not serve only to confirm a law from which practical predictions can be made. Its ideal is rather to understand the phenomenon itself in its unique and historical concreteness.[34]

I am sure Gadamer did not have anything like the lady's hat in mind, but "the phenomenon in all its unique and historical concreteness" is wonderfully illustrated by Wisdom's Taj Mahal. Aristotle would also have insisted that in an argument, a good example should not only be familiar to the reader, and thus shared, but also be novel, or better, *vivid*. Here I like Hume's word *vivacity*: that is what strikes the human mind in debate and much else.

In practical fields we grasp particular facts of experience more clearly, and have more certainty of their truth, than we ever do about the general principles that we may use to account for them.[35]

Back to Kuhn's Paradigms

I began by saying, with Daston, that no one has given a satisfactory analysis of Kuhn's idea of a paradigm as example—and that despite the fact that he said it was the key new idea of *Structure*. But I suggest that it is not only Kuhn's idea that we have failed to explicate with clarity and precision. The difficulty is pervasive in the family of notions we variously call analogy, models, similarity, likeness, resemblance. This is not because they are obscure but because they are relational; what is enlightening to one audience is often devoid of use or content to another. I have suggested that this fact dogged the use of examples from their introduction into philosophy with Aristotle's *paradeigma* in the rhetoric. I have sketched a little history of ups and downs of argument by example, not for its sake as flimsy history but as a way to illustrate the persistence of the difficulties. I do not fault Kuhn for not bequeathing us a clear concept of paradigm. I praise him for giving new fire to it. His was a brilliantly novel use of an ancient idea.

Disclosure

I learned about the use of examples in philosophy in the very first course of lectures in philosophy that I ever attended, given by John Wisdom in the Michaelmas term at Cambridge University, 1956. I am inclined to quote Father William: "The muscular strength which it gave to my jaw has lasted the rest of my life." Hence, I was well primed when, in the late 1960s, teaching in East Africa, I noticed how good the undergraduates were at case-by-case arguments. One of my pupils, Cyprian Bafutwabo, introduced me to the use of proverbs in traditional argument and law among his peoples, as I have described elsewhere.[36] This was a ritualized use of a stock of about five thousand paradigms. Bafutwabo's father or uncle had been the primary informant for Rodagem.[37]

The font of the interest in example was of course Wittgenstein, who insisted that it did not matter whether his examples were drawn from real life, or were fictional cases. *Enlightening* ones, for sure.

Stephen Toulmin's work on argument and casuistry descends as strictly from this tradition as my own thoughts, though he seldom acknowledges Wisdom.[38] Geoffrey Lloyd, who may have attended to Aristotle on example more closely than another recent classical scholar, worked in this philosophical milieu.

As for myself, I can best quote hearsay. "As Ian Hacking is reputed to have once said, 'Philosophy is the art of the good example.'"[39] Whether I actually said that, I endorse the sentiment, though customary caution would require some caveats. I have always insisted that there are many ways to do philosophy. On an even more personal note, I am accurately quoted by Madsen, Servan and Øyen: "I am a philosopher of the particular case."[40] But unlike Wittgenstein, I am fascinated by real particulars, which I usually find stranger—or at any rate more *enlightening*—than imagined ones.

Notes

1. Ian Hacking, "Introductory Essay," in *The Structure of Scientific Revolutions*, 4th ed., vii–xxvi (Chicago: University of Chicago Press, 1992).
2. Kuhn, "Postscript—1969," in *The Structure of Scientific Revolutions*, 2nd ed. (Chicago: University of Chicago Press, 1970), 186.
3. John Arthos, "Where There Are No Rules or Systems to Guide Us: Argument from Example in a Hermeneutic Rhetoric," *Quarterly Journal of Speech* 89 (2003): 320–34 (quote, 320).
4. Masterman, Margaret. "The Nature of a Paradigm," in *Criticism and the Growth of Knowledge*, ed. Imre Lakatos and Alan Musgrave, 59–89 (Cambridge: Cambridge University Press, 1970).
5. Rom Harré, "Where Models and Analogies Really Count," *International Studies in the Philosophy of Science* 2 (1988): 118–33.
6. Thomas S. Kuhn, "Second Thoughts on Paradigms," in *The Structure of Scientific Theories*, edited by Frederick Suppe, 459–82 (Urbana: University of Illinois Press, 1974).
7. David Bloor, "Left and Right Wittgensteinians," in *Science as Practice and Culture*, edited by Andrew Pickering, 266–83 (Chicago: Chicago University Press, 1992). For what many will judge to be an overly critical survey of the literature, see David Stern,

"Sociology of Science, Rule-following and Forms of Life," in *History and Philosophy of Science. New Trends and Perspectives*, ed. Michael Heidelberger and Friedrich Staedler, 347–67 (Dordrecht: Kluwer, 2002).

8. Ludwig Wittgenstein, *Philosophical Investigations*, trans. G. E. M. Anscombe, 3rd rev. ed. (Oxford: Blackwell, 2001), § 1–9.

9. Aristotle, *Rhetoric* 1.1. 1354ª20. G. A. Kennedy, *Aristotle: On Rhetoric. A Theory of Civic Discourse.* 2nd ed. (New York: Oxford University Press, 2007), 31.

10. Myles Burnyeat, "Enthymeme: Aristotle on the Rationality of Rhetoric," in *Essays on Aristotle's Rhetoric*, ed. A. Rorty, 88–115 (Berkeley: University of California Press, 1996).

11. Aristotle *Rhetoric* I.1 1354ª 1–11; Kennedy, *Aristotle: On Rhetoric*, 30 (his square brackets).

12. Aristotle *Rhetoric* 1356ᵇ; Kennedy, *Aristotle: On Rhetoric*, 40.

13. Myles Burnyeat, "Enthymeme: Aristotle on the Logic of Persuasion," in *Aristotle's Rhetoric: Philosophical Essays*, ed. D. Furley and A. Nehamas, 3–55 (Princeton, NJ: Princeton University Press, 1994); Burnyeat, "Enthymeme: Aristotle on the Rationality of Rhetoric."

14. Arthos, "Where There Are No Rules or Systems to Guide Us," 322.

15. Burnyeat, "Enthymeme: Aristotle on the Rationality of Rhetoric"; Ian Hacking, *The Emergence of Probability* (Cambridge: Cambridge University Press, 1975); Geoffrey Lloyd, *Polarity and Analogy: Two Types of Argumentation in Early Greek Thought* (Cambridge: Cambridge University Press, 1966); Lloyd, *Ancient Worlds, Modern Reflections: Philosophical Perspectives on Greek and Chinese Science and Culture* (Oxford: Clarendon Press, 2004).

16. Aristotle *Topics* VIII 156ᵇ10–16.

17. Modern readers may not realize how poignant the example is. Mill published his *System of Logic* in 1843. The very conservative Duke of Wellington had been a British hero since Waterloo, in 1815. But there was hope for liberals such as Mill: the duke, too, was mortal. Mill, as is well known, went on to argue that every inductive syllogism was a *petitio principii*, begging the question, because the major premise, about the genus, was never fully established and assumed the conclusion. And so he went on to develop his own remarkable theory about induction. Aristotle, in contrast, would have emphasized the role of getting the "right" genus, an approach that the antiessentialist Mill rejected out of hand.

18. Aristotle *Rhetoric* 1357ᵇ–1358ª.

19. Aristotle *Rhetoric* 1392ᵇ–1393ª.

20. Aristotle *Rhetoric* II 20; Kennedy, *Aristotle: On Rhetoric*, 162.

21. Kennedy, *Aristotle: On Rhetoric*, 163.

22. Aristotle, *Rhetoric*, 1393ᵇ6: Kennedy, *Aristotle: On Rhetoric*, 180.

23. Aristotle, *Rhetoric*, 1394ª7-8: Kennedy, *Aristotle: On Rhetoric*, 181.

24. Myles Burnyeat, "The Origins of Non-Deductive Inference," in *Science and Speculation: Studies in Hellenistic Theory and Practice*, ed. J. Barnes et al., 193–238 (Cambridge: Cambridge University Press, 1982), 234.

25. J. D. Lyons, *Exemplum: The Rhetoric of Example in Early Modern France and Italy* (Princeton, NJ: Princeton University Press, 1990).

26. François Rigolet, "The Renaissance Crisis of Exemplarity," *Journal of the History of Ideas* 59 (1998): 557–63.

27. Montaigne, *The Essays of Montaigne*, trans. E. J. Teichman (London: Oxford University Press, 1927), 546 (*Essays* III ch. 13; emphasis added).

28. Nelson Goodman, "Seven Strictures on Similarity," in *Problems and Projects* (Indianapolis, IN: Bobbs-Merrill, 1972).
29. John Wisdom, "The Logic of God," in *Paradox and Discovery* (Oxford: Blackwell, 1965), 2.
30. Ibid., 3.
31. Ibid., 4.
32. Cora Diamond, *The Realistic Spirit: Wittgenstein, Philosophy, and the Mind* (Cambridge, MA: MIT Press, 1991), 28.
33. Arthos, "Where There Are No Rules or Systems to Guide Us," 322.
34. Hans-Georg Gadamer, *Truth and Method*, trans. G. Barden and J. Cumming (London: Sheed and Ward, 1975), 4–5.
35. Albert R. Jonsen and Stephen Toulmin, *The Abuse of Casuistry: A History of Moral Reasoning* (Berkeley: University of California Press, 1988), 25–26.
36. Ian Hacking, "What Logic Did to Rhetoric." *Journal of Cognition and Culture* 13 (2013): 419.
37. François-Marie Rodagem, "Sagesse Kirundi; proverbs, dictons, locutions usités au Burundi," in *Annales du Musée Royal du Congo Belge. Série in 8o*, vol. 34 (1961). *Sciences de l'Homme*. Expanded in *Paroles de sagesse au Burundi* (1961; Leuven: Peters, 1983).
38. Stephen Toulmin, *The Uses of Argument* (Cambridge: Cambridge University Press, 1958); Jonsen and Toulmin, *The Abuse of Casuistry*.
39. Daryn Lehoux, *What Did the Romans Know?* (Chicago: Chicago University Press, 2011), 87.
40. O. J. Madsen, J. Servan, and S. A. Øyen, "'I Am a Philosopher of the Particular Case': An Interview with Holberg Prize Winner 2009 Ian Hacking," *History of the Human Sciences* 26 (2013): 32–51 (quote, 32, 39).

Kuhn in his Berkeley home from the period of his tenure
on the University of California faculty

SIX

History of Science without *Structure*

LORRAINE DASTON

Prelude: In August 1950, Professor Theodor Erismann, director of the Institut für Experimentelle Psychologie of the Universität Innsbruck in Austria, conducts a ten-day experiment with inverting glasses, enlisting his young colleague (identified as Dr. M.) as subject. The inverting glasses use mirrors rather than the heavier and more unwieldy prismatic lenses of earlier experiments and make it possible, though not comfortable, for subjects to wear the glasses for longer periods, in this case ten days (figure 6.1). At first Dr. M. is disoriented and unable to navigate even familiar environments. Professor Erismann equips him with a blind person's arm badge and stick and guides him by the arm. For at least the first few hours of the experiment, Dr. M. is effectively blind.

Introduction: Structure

I begin with an obvious and, for Thomas Kuhn, highly suggestive example of a structural transformation: an experiment on the psychology of perception in which the normal visual field is turned upside down by inverting glasses and then eventually rights itself as the subject becomes accustomed to the upside-down perspective. For Kuhn, experiments with inverting glasses served as a powerful metaphor (and perhaps more than a metaphor) for paradigm shifts, and in particular, for what was wrong with the view that theory change was simply a matter of reinterpreting the same, stable perceptual data: "Rather than being an interpreter, the scientist who embraces a new paradigm is the like the man wearing inverting lenses. Confronting the same constellation of objects as before and knowing that he does so, he nevertheless finds them transformed through and through in many of their details."[1] I will return to such experiments and

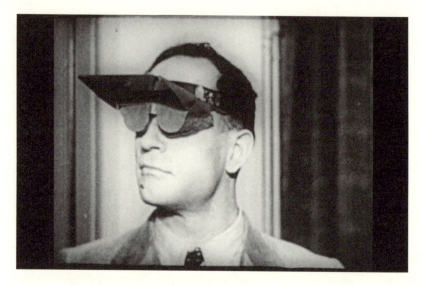

Figure 6.1. Dr. M. puts on the inverting glasses. (Still [01.02] from
Theodor Erismann and Ivo Kohler, *Die Umkehrbrille und das aufrechte Sehen*
[Institut für Experimentelle Psychologie, Universität Innsbruck, 1950].)

the use Kuhn made of them intermittently in an attempt to make sense of
structure—both for himself and perhaps also for us.

"Structure" was a word to conjure with in 1962. Claude Lévi-Strauss's
Anthropologie structurale had appeared in 1958 (translated into English in
1963); Noam Chomsky's *Syntactic Structures* was first published in 1957
and had reached its fifth printing by 1965. Even if they didn't brandish the
word "structure" in their titles, a cluster of other influential books in the
humanities and social sciences published circa 1960 raised hopes that the
complexities of, say, the plays of Racine or cultural taboos or bargaining
might reveal simpler basic structures in the way that an X-ray revealed skel-
etons.[2] The runaway success of Thomas Kuhn's *The Structure of Scientific
Revolutions* did its part to glamorize an already up-and-coming word.

Yet probably no word strikes today's historians of science reading Kuhn
(if they do) as more dusty and dated than the once glittering "structure."
This is not because the whole book is a fossil from a bygone era. Even if
it is no longer assigned in courses across the university, much of Kuhn's
analysis still seems fresh and even avant garde: the close studies of scientific pedagogy that he flagged as crucial to understanding the cognitive
and social cohesion of research communities are still a desideratum; topics such as the know-how implicit in mastering scientific paradigms have

been revived by the history of the body and other explorations of what is often called, not always accurately, tacit knowledge. Even though many of the polestar words that do now guide the history of science—"context," "controversy," "consensus"—were first made luminous in Kuhn's *Structure*, "structure" itself has lost its shine.

The fall of "structure" contrasts with the inexorable rise of "scientific revolutions," the other part of Kuhn's title. His determination to wrest the definite article and majuscules away from "*the* Scientific Revolution" of the seventeenth century and to apply the minuscule version to many other episodes in the history of science has succeeded beyond his wildest expectations. We now nonchalantly speak of all manner of revolutions— the Darwinian, the Einsteinian, even the genomic—even if we balk at the now-vulgar "paradigm." Kuhn's central contention, namely, the inevitability of recurrent scientific revolutions in all fields, in the emphatic plural of his title, has become a commonplace.[3]

So why has "structure" fared so badly, at least within the history of science? Not because Kuhn's own version of structure has been decisively refuted, much less because an alternative pattern of historical development has replaced it. Most historians of science no longer believe that *any* kind of structure could possibly do justice to their subject matter. The very idea of looking for overarching regularities in the history of science seems bizarre, a kind of leftover Hegelianism seeking a hidden, inexorable logic in the apparent vagaries of history—in Kuhn's case, the last attempt to give Reason (now incarnate in science) a rational history. The words that have replaced "structure" among the bywords of the humanities and social sciences—terms like "culture" and "context" and "thick description"— deliberately baffle all attempts at generalization, whether philosophical or sociological, that aspire to span places and periods. Since roughly the 1990s, the focus in these disciplines has shifted from the streamlined to the dense and detailed; the professed aim has been to "complexify" rather than simplify and to reveal variability rather than uniformity. Even the adjectives routinely used to praise lectures index this shift in intellectual sensibility from Bauhaus (or perhaps 1960s Swedish modern) to Baroque: good papers are "rich," no longer "acute" or "incisive."

Kuhn himself heralded this change, perhaps unintentionally and almost certainly not realizing that it would undermine a search for structures in the history of science. In a sharp and occasionally bitter 1971 article on why historians had not embraced the history of science, Kuhn placed the blame squarely on the shoulders of the historians.[4] Yet he was unwavering in his conviction that the future of the history of science lay in history departments. By rejecting the "Whiggishness" inherited from scientists' own version of their history, the history of science had "become potentially a fully historical enterprise."[5] It would become historicist, in the sense of situating science in its specific context, just as the history of art (often a point of reference for Kuhn) embedded styles and genres in particular periods and places.

That potential has been fulfilled, perhaps with a vengeance. History of science has never been more resolutely historical in its methods (archival) and modes of explanation (contextual); many if not most of its practitioners teach in history departments, where Kuhn thought they belonged. Most strikingly, the historicist program in the history of science has fractured the once-monolithic "science" into the sciences and raised serious questions as to whether the term can be applied at all to the premodern epoch.[6] Yet the historicism Kuhn prophesied and welcomed has ultimately dismantled the structures he sought: an essential tension at the heart of his own still riveting vision for the history of science.

In this essay I would like to explore the legacy of Kuhn's tension between structure and historicism for the history of science now. I will argue three theses: first, that historicism has triumphed so completely over structures that the history of science may soon dissolve its own subject matter; second, that abandoning structure also meant abandoning close ties with

Figure 6.2. Dr. M. confusedly turns his cup upside down under the water pitcher. (Still [03.47] from Erismann and Kohler, *Die Umkehrbrille*.)

both the philosophy and sociology of science, at least in the short term; and third, that there is nonetheless considerable potential in at least one of Kuhn's structures to reconnect even a thoroughly historicized history of science with these once-kindred fields. But it may come at the cost of rethinking just what a structure is.

Interlude: During the first few days of the experiment, Dr. M. is confused and clumsy. He consistently confuses up and down directions, reaching downward to catch a balloon wafting upward and inverting his cup to catch a stream of water (figure 6.2).

History, the Universal Solvent

The battle lines that divided the history of science into warring factions in the 1970s and '80s have all but disappeared in the current literature of the field. Ask younger scholars whether science is a social construction or an approximation of reality, and their response is likely to be "both," accompanied by a condescending smile at your quaint phrasing. Try to organize a session titled "Internal versus External Approaches to the History of Science" at the annual meeting of the History of Science Society or some other major professional gathering, and, faster than the speed of electrons, you will receive a rejection note with the explanation that space on the program is at a premium and reserved for more topical themes. The words that once provoked shouting matches—"incommensurability," "irrationality," "relativism," and all the other ghouls and goblins allegedly let loose from the Pandora's box that was *The Structure of Scientific Revolutions*—all these now elicit barely a yawn, at least among the book's primary intended audience of historians of science.

To explain this precipitous drop in disciplinary blood pressure would require a book in its own right, but the short version is that historians of science became historians. This is what Kuhn had predicted and preached: historians of science must free themselves from the teleological narratives of philosophers; they must understand past science in the past's own terms through "deep and sympathetic immersion" in the sources;[7] they must correct the "depreciation of historical fact" engrained in the "ideology of the scientific profession;"[8] they must seek employment in history departments.

And lo, all this came to pass, but as is so often the case with prophecies, not quite as the prophet himself imagined it would. Historians of science did indeed immerse themselves deeply and sympathetically in primary sources; they adopted the historians' *rite de passage* and baptized their

Ph.D. students in the archives; they forsook the anachronism of scientific textbooks; they even got themselves hired by history departments. Yet the end result of all this historicizing was to unravel much of the structure of Kuhn's own *Structure*, including what he believed to be the *sine qua non* of "mature" science, the quality that distinguished it from otherwise similar human activities such as art or politics or even the still immature social sciences—namely, "the unparalleled insulation of mature scientific communities from the demands of the laity and of everyday life."[9] In order to become a science, a field of inquiry had to become, in Kuhn's words, "esoteric," judged only by its own standards and those standards fixed by the puzzle-solving exemplars of the reigning paradigms. Although Kuhn's own career, from his first choice of physics to his second choice of history of science, was molded by the most famous science-on-commission in human history, the atomic bomb project at Los Alamos,[10] he insisted on the autonomy (or perhaps autarchy) of science as its defining characteristic. This premise of "unparalleled insulation" was precisely what the newly historicized history of science challenged, starting in the late 1980s and 1990s. The "Science in Context" program, which also lent its name to a journal, was dedicated to showing that even science was part and parcel of its historical time and place and subject to the same contingencies as other human activities.

Although the old externalist and the new contextual approaches to the history of science at first appeared to march under the same banner, they eventually diverged sharply. Externalist history of science was just that: it investigated the forces—institutions, politics, religion, even philosophy—supposedly "outside" science, thereby accepting the premise that it made sense to talk about science having an "inside" and an "outside." Moreover, its focus was on structural factors, Marxist means of production or Weberian religious ideologies, just as coeval social history looked to economics and sociology to supply its analytic categories. In contrast, contextual history of science took its cue from cultural history and the richly detailed, circumstantial genre of the microhistory. Its cognate social science was not economics but anthropology; its motto was "all knowledge is local knowledge." "Context" chimed with "texture," and contextual history of science aspired to narratives as intricately woven as tapestries rather than structures as firm as steel girders.

At the center of contextual history of science were concrete practices: what scientists did rather than what they said. Practices could embrace everything from the fine adjustments required for a precision measurement to collecting natural history specimens to computer simulations. Although

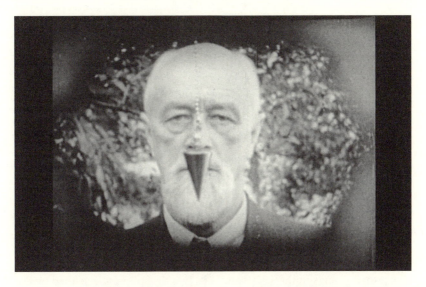

Figure 6.3. Dr. M. turns a face right-side-up by dangling a plumb bob in front of it.
(Still [07.47] from Erismann and Kohler, *Die Umkehrbrille.*)

no one could dispute that such practices were integral and therefore "internal" to science, the pursuit of their historical origins often led to "external" sites. The techniques of the laboratory experiment led back to the tests of the artisan's workshop; the journal of scientific observations piggybacked on the Renaissance humanist's commonplace book; graphical data display borrowed from methods used by engineers to monitor steam engine performance. Practices broke down the wall erected between internal and external approaches to the history of science and made the wars waged between proponents of each all but incomprehensible to the next generation of scholars.

Interlude: Now three to five days into the trial, Dr. M. can selectively right his vision by using his sense of touch or background knowledge about how heavy objects fall downward. By dangling a plumb bob in front of a face, he can make it turn right-side-up, even if the rest of his visual field still seems inverted (figure 6.3).

Farewell to Sociology and Philosophy

The impact of contextualized, practice-centered approaches to the history of science was simultaneously to make the history of science more like general history, just as Kuhn had hoped, but at the high cost of weakening its ties to sociology and philosophy of science, and indeed, to science itself. For the scientists, a history of science no longer written from the standpoint of the present was like a performance of *Hamlet* without the prince; for the sociologists and philosophers, a history of science crammed with irreducibly local details and contingencies could no longer serve as a treasury of examples from which generalizations about the nature of science and scientists might be minted. For the decades following the publication of *The Structure of Scientific Revolutions* in 1962, the history of science had been arguably the most highly theorized specialty within all of history; now it is for the most part only the history of very recent science that still engages the attentions of sociologists, anthropologists, and philosophers, usually under the capacious roof of science studies.

This state of affairs would not only have surprised Kuhn; it would also have flummoxed the post-Kuhnian generation of historians of science who began their careers in the 1970s and '80s in a discipline still defined by the debates ignited by Kuhn's book. Not only *dernier cri* science of the past few decades but also that of the Scientific Revolution and the Enlightenment was fodder for science studies approaches. Historians of early modern as well as modern science were in intense and often fruitful dialogue with the Edinburgh and Bath schools of the sociology of science and made common cause with philosophers of science who wanted to understand the nature of science on the basis of what scientists actually practiced, not just what they preached in methodological prefaces. Every graduate student in the history of science could explain the strong programme and experimenter's regress; historians, sociologists, and philosophers all contributed to a brilliant burst of studies on scientific experiment.[11] Until the mid-1990s, articles devoted to science studies topics appeared regularly in *Isis*, and *Social Studies of Science*, the flagship journal of science studies, returned the favor by publishing articles dealing with pre-1900 topics. In 1996, however, the latter journal introduced an "occasional series of 'Historic Papers,'" under which rubric only three items appeared from 1996 to 2010. Conversely, the "Focus" section of *Isis*, initiated in 2004 to spotlight topics of overarching interest in the history of science, has featured only a few themes with strong science studies affinities.

These developments chart the shifting position of the history of science within the map of the disciplines: ever closer to history and ever farther from philosophy, sociology, and anthropology. It is not a coincidence that the arc of the influence of Kuhn's *Structure* and the controversies it inspired declined steeply at exactly the same time. There was, to use one of Kuhn's

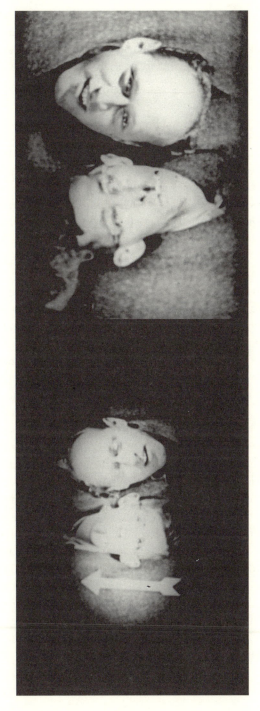

Figure 6.4. Dr. M. sees two heads that are simultaneously oriented in opposite directions but both of which appear upright, an experience that cannot be depicted in the film but only suggested by the arrow. (Still [0.909] from Erismann and Kohler, *Die Umkehrbrille*.)

own memorable phrases, an essential tension between the two parts of the book's program: on the one hand, the search for structures (the part that fired the imagination of the philosophers and sociologists); and on the other, the aim to historicize the history of science (the part that led to a fusion with general history).

Interlude: Dr. M.'s use of his background knowledge (e.g., that flames and smoke point upward) to selectively rectify his vision can lead to paradoxical vision. He is shown two heads (with bodies hidden) oriented in opposite directions. The apparently upside-down head then begins to smoke a cigarette. The flame and drifting smoke prompt Dr. M. visually to invert that head so that it is also right-side-up. Yet he still sees the heads as oriented in opposite directions (figure 6.4).

The Soft Center of Structure

Kuhn would probably have been appalled at this outcome, even if it ultimately resulted from his own historicism. He was vehemently convinced of the integrity and distinctiveness of science as a way of knowing and as a form of communal life, both of which centered on paradigms. All of the most structured parts of *Structure* relate to paradigms: how disciplines became "mature sciences" with the advent of their first paradigm; how paradigms "provide scientists not only with a map but also with some of the directions for map-making"; how cumulative progress is only possible within a paradigm; why it was impossible for paradigms to peacefully coexist, for "proponents of competing paradigms practice their trades in different worlds." Only "orthodox theology" could compete with the "narrowness and rigidity" of scientific education into the reigning paradigm. [12] The history of science was nothing more or less than the history of paradigms as they rose and fell like empires. According to Kuhn, paradigms could even mold perceptions, encoding knowledge into the very neural processes of scientists. [13]

This brings me back to those experiments with inverting glasses. Kuhn played with the idea that such radical shifts in perception might be not just an analogy to but an actual model of what happens when paradigms collide. When Galileo saw a pendulum where Aristotle had seen constrained fall, when Lavoisier saw oxygen where Priestley had seen dephlogisticated air—perhaps, Kuhn hesitantly suggested, these paradigm transitions were in some expansive sense of "seeing" equivalent to putting on inverting glasses? "Did these men really see different things when looking at the same sorts of objects? Is there any legitimate sense in which we can say that they pursued their research in different worlds?"[14] Kuhn returned again and again to perception as a way of capturing the irreversible, all-at-once, all-enveloping quality of paradigm shifts. Whatever kind of structures paradigms were, they were at once self-evident, thorough, and sudden. "Scientists then often speak of 'the scales falling from the eyes' or of the 'lightning flash' that 'inundates' a previously obscure puzzle, enabling its components to be seen in a new way that permits its solution."[15]

Yet if you've been attending to the images and interludes describing the actual experiment performed with inverting glasses, you'll have noticed significant disanalogies with Kuhn's paradigm shifts. For one thing, the transition is gradual and piecemeal: Dr. M. was able to rectify certain parts of his visual field, but not others, after a few days. For another, he could selectively effect the transition by using his other senses or knowledge about the properties of how certain objects in the world behave, like the plumb bob. Finally, and most disconcertingly for Dr. M. and, by implication, for Kuhn, he could straddle upright and inverse vision, as in the case of the paradoxical heads. Unlike Kuhn's scientists, he could simultaneously inhabit two visual worlds.

I do not draw attention to these disparities in order to discredit Kuhn's paradigms or even the analogy he repeatedly drew between inhabiting a paradigm and seeing the world. Rather, I want to point out that Kuhn's implicit view of what a structure had to be—as confining as a straitjacket, as all-pervasive as the world, as irrefragable as perception—and therefore of how such structures must be transformed—as suddenly as a lightning flash—correspond neither to perception nor to the most interesting aspect of his own account of paradigms.

Although, on Kuhn's account, paradigms structured science just as their succession structured the history of science; they were soft, mushy even, at their cores. I am not referring to the twenty-one different senses in which Kuhn used paradigms, famously cataloged by philosopher Margaret Masterman.[16] Rather, I mean the one sense of paradigm that Kuhn himself consistently underscored as the most important: paradigms as exemplars, as opposed to sets of rules. In his 1969 postscript to *Structure*, he described this sense of paradigm as "models or examples, [that] can replace explicit rules as a basis for the solution of the remaining puzzles of normal science" as philosophically "deeper" than the others,[17] even though he was at a loss to explain exactly how it worked. He admitted that the neural mechanisms ultimately responsible for scientific learning and problem-solving must be governed by laws but nonetheless maintained that the conscious and unconscious processes by which scientists mastered and extended paradigms could not be reduced to rules. Rightly anticipating charges of irrationalism and mysticism, he stoutly defended the knowledge transmitted by paradigms as genuine knowledge: "When I speak of knowledge embedded in shared exemplars, I am not referring to a mode of knowing that is less systematic or less analyzable than knowledge embedded in rules, laws, or criteria of identification."[18]

Yet to date, neither Kuhn nor anyone else has succeeded in hammering out a systematic, analytical language for talking about knowledge without rules. The ways in which Kuhn and others have instead circumscribed the problem only thicken the murkiness that surrounds it. Appeals to "intuition," "tacit knowledge," or "conversion" have only strengthened the impression—the very opposite of the one Kuhn meant to convey—that knowledge without rules is no knowledge at all, closer to gut feelings than cognition. It should be pointed out that Kuhn's conundrum challenged other disciplines besides the history of science. Philosophers impressed by Wittgenstein's demonstration of the inadequacy of rules to account for language usage or social conduct[19] have also been unable to go beyond Wittgenstein's gesture toward "practice," a term akin to and just as opaque as Michael Polanyi's "tacit knowledge."[20] Cognitive scientists intent on modeling all mental processes by algorithmic rules have been similarly stymied by the broad set of human capacities loosely referred to as "pattern recognition." Legal scholars have been unable to formulate rules for the application of precedents (or even laws) to new cases tried by the courts. When rules fail, judgment is invoked—but the term is now almost always modified by the damning adjective "subjective," a sure sign that we have left the crystalline realm of the rational.

Or so it would seem. For the most diverse reasons, ranging from the wildfire spread of computers and invention of algorithms for almost everything to the expansion of regulatory bureaucracies at both national and international levels, rules dominate our conceptions of order and rationality. This is perhaps the last bastion of the otherwise outmoded "structures" of the 1960s. The very word "rationality" is subtly distinguished from the cognate "reason" by the assumption that the former can be reduced to rules, preferably algorithmic rules, while the latter must invoke judgment and self-conscious deliberation. Increasingly, rationality has enjoyed the upper hand. This is why Kuhn's non–rule-governed reasoning from exemplars strikes us at best as just so much hand waving and at worst as a descent into a swamp of subjectivity.

But this conviction that only rules can offer the analytic clarity, reliability, and objectivity demanded of all kinds of reasoning worthy of the name, including scientific reasoning, is itself a product of history, and quite recent history at that. It is startling to realize that in ancient Greek, the words for "rule" (*canon*) and "paradigm" or "example" (*paradeigma*)[21] were often used as synonyms, a usage perpetuated in Latin and in modern European vernaculars until at least the late eighteenth century.[22] Here, for example, is Pliny the Elder (23–79 CE) upholding the Greek sculptor Polykleitos's statue *Doryphoros* (*The Spear-Bearer*) as the *canona* (the Latinized version of the Greek word, although the usual Latin term *regula* had a similar field of associations[23]), the model of male beauty worthy of imitation by all artists. "He also made what artists call a 'Canon' [*quem canones artifices vocant*] or 'Model Statue,' as they draw their artistic outlines from it as from a sort of standard."[24] Or, fast-forwarding almost two thousand years to Enlightenment France, the *Encyclopédie*'s sample sentence for its first definition of the entry "*Règle, Modèle*" is "The life of Our Savior is the *rule* or the *model* for Christians."[25] In both ancient Greek and Latin grammar, the word *canon*, or *regula*, was used along with *paradigma* to denote that paradigm of paradigms, namely, patterns of inflections such as verb conjugations intoned by schoolchildren over the centuries: *amo, amas, amat, amamus, amatis, amant*. Still more confusingly, at least for modern sensibilities, is the fact that these premodern usages of words for "rule" also include our more exact and rigid sense. The ancient Greek word *canon*, for example, denoted painstaking exactitude, especially in connection with the arts of building and carpentry, but also in a figurative sense when applied to other domains such as art, politics, music, and astronomy (astronomical tables were called *canones*). According to Galen, the same Polykleitos who fashioned the *Doryphoros* statue was the author of a lost treatise entitled *Canon* in which he allegedly specified the exact proportions of the human body to be followed by artists; such prescriptive measurements of classical statues were still on display in the eighteenth century.[26] This cluster of meanings evokes the rigor of mathematics, both as the geometric doctrine of proportions and as tool of measurement and computation—meanings that happily coexisted with the cluster centered on models and paradigms.

For Kuhn, and indeed, for most of his readers, such usages would have been profoundly puzzling. The paradigm case, so to speak, of a modern rule is an algorithm, defined by mathematicians as definite, general, and conclusive.[27] Yet the fertility of Kuhn's paradigms for scientific practice lay precisely in their being indefinite enough to point by analogy to kindred cases, specific enough to anchor problem-solving techniques, and open-ended enough to support research programs.[28] The fact that the premodern usage of rule could embrace all of these meanings—to Kuhn's mind, the equivalent of meaning A and not-A simultaneously—does not in itself resolve the modern tension between rule and paradigm. But it does reframe the problem: instead of asking, as Kuhn felt obliged to, how knowledge that cannot be reduced to rules can nonetheless be considered systematic knowledge worthy of the name, we might instead ask, first, how the meaning of rule has so radically narrowed since the mid-nineteenth century, and second, why the narrowed meaning of rule as algorithm has become normative in so many disciplines, from philosophy to cognitive science to economics. Yet these developments, although widespread and deep cutting, have encountered obstacles. The stubborn resistance of phenomena as diverse as legal judgments, language acquisition, pattern recognition, and scientific training to rule-governed accounts, despite the best efforts of the best scholars for almost a century now, suggests that the phenomenon of learning and reasoning from exemplars that Kuhn identified as the heart of scientific paradigms is both real and encompasses far more than just science.

From the standpoint of thoroughly historicized history of science, this breadth is a positive advantage. Kuhn thought paradigms erected walls around science and gave it its "esoteric" and singular character. This may still be arguable for specific examples, such as Newtonian physical optics or genetics after Watson and Crick. But the form of reasoning Kuhn described as paradigmatic for paradigms, reasoning from exemplars, crops up everywhere—in the history of knowledge more broadly construed (e.g., ethnobotany) and scholarship (e.g., classical philology), as well as in the history of science, narrowly construed. Historians of science who no longer share Kuhn's convictions about the singularity of science could embrace it without compunction. Moreover, as in the experiment with the inverting glasses, learning how to recognize and reason with exemplars is a gradual process that proceeds in fits and starts, neither a thunderbolt intuition nor a conversion experience. Learning unfolds in time, can be guided by pedagogy, and is context-sensitive—and therefore can be studied by historians. Philosophers, sociologists, and anthropologists of science would for their part have a problem worthy of their mettle, the solution of which would both explain the extraordinary solidarity of communities united by shared habits of perception and thought (Ludwik Fleck's *Denkkollektiven*)[29] as well as rehabilitate judgment as a faculty that can be analyzed, even if it cannot be reduced to rigid rules.

Interlude: By the seventh or eighth day of the experiment, Dr. M. is able to navigate his environment effortlessly without any assistance. His visual field now appears "normal" to him (figure 6.5).

Figure 6.5. Dr. M. riding his bicycle with assurance through crowded streets. (Still [09.58] from Erismann and Kohler, *Die Umkehrbrille.*)

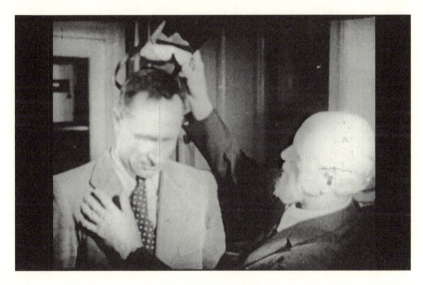

Figure 6.6. Dr. M. removes the inverting glasses after ten days. (Still [11.48] from Erismann and Kohler, *Die Umkehrbrille.*)

Admittedly, this is a tall order, and some might be inclined to say that it is as hopeless as trying to detect a backbone in a jellyfish. But this would be to assume what has yet to be proven, despite decades of effort: namely, that reasoning from exemplars can indeed be cashed out as rule-following. Kuhn insisted that it could not, and even if the structures he envisioned for the history of science have melted away like mirages, his final conception of paradigm might yet yield a structure capacious enough to bring the history, philosophy, and sociology of science—and much else—back together again.

Postscript: After ten days, the inverting glasses are removed. At first, Dr. M.'s vision is once again inverted; suddenly, everything seems upside-down again. But after a few minutes, his vision rights itself: the world back in order (figure 6.6).

Notes

Some of the material in the introduction to this article was published in Lorraine Daston, "Structure," *Historical Studies in the Natural Sciences* 42(2012): 496–99. I thank the University of California Press for permission to include it here.

1.   Thomas S. Kuhn, *The Structure of Scientific Revolutions*, 4th ed. (Chicago: University of Chicago Press, 2012), 121–22. Kuhn references the early experiments done by the Leipzig-trained Berkeley psychologist George Stratton and a 1935 survey article of subsequent experiments with inverting lenses (113n1). However, Stratton's experiment was for monocular vision only; he used tubes fitted with two convex lenses and found the parallax problems of coordinating two eyes "too severe" (George M. Stratton, "Some Preliminary Experiments on Vision without Inversion of the Retinal Image," *Psychological Review* 3[1896]: 611–17, on 612). Theodor Erismann began using lighter, better-coordinated mirror glasses in the 1920s and 1930s, which made longer, binocular experiments possible, although subjects sometimes had to break off the trials because of "dizziness, nausea, depression, or painful pressure on nose and ears" (Ivo Kohler, *Über Aufbau und Wandlungen der Wahrnehmungswelt., insbesondere über 'bedingte Empfindungen'*, Österreichische Akademie der Wissenschaften, Philosophisch-historische Klasse, vol. 227 [Vienna: Rudolf M. Rohrer, 1951], 14). Kohler gives an overview of all previous research on inverted vision and describes the experiments that were the subject of the documentary film on 18–19. Kuhn does not seem to have known of the work of Erismann and Kohler. The film can be seen at https://www.youtube.com/watch?v=z1HYcN7f9N4 (accessed March 15, 2015).

2.   Among the other iconic works in this vein and of this era were Roland Barthes, *Sur Racine* (Paris: Editions du Seuil, 1960), Mary Douglas, *Purity and Danger: An Analysis of Concepts of Pollution and Taboo* (New York: Praeger, 1966), and Thomas C. Schelling, *The Strategy of Conflict* (Cambridge, MA: Harvard University Press, 1960).

3.   Kuhn himself thought this point would encounter so much resistance among readers conditioned to think of scientific development as smoothly continuous that he

devoted an entire chapter to explaining it (Kuhn, "The Invisibility of Revolutions," chapter 11 in *Structure*, 135–42).

4. Thomas S. Kuhn, "The Relations between History and the History of Science," *Daedalus* 100 (1971): 271–304, repr. in Kuhn, *The Essential Tension: Selected Studies in Scientific Tradition and Change* (Chicago: University of Chicago Press, 1977), 127–61, on 154.

5. Ibid., 150.

6. These issues continue to exercise historians of science: see, e.g., Jan Golinski, "Is It Time to Forget Science? Reflections on Singular Science and Its History," *Osiris* 27 (2012): 19–36; and Peter Dear, "Science Is Dead; Long Live Science," *Osiris* 27 (2012): 37–55.

7. Thomas S. Kuhn, "History of Science," *International Encyclopedia of Social Sciences* (New York: Free Press, 1972), 13: 74–83, on 77.

8. Kuhn, *Structure*, 138.

9. Ibid., 163.

10. Kuhn encountered the history of science at Harvard through James Bryant Conant's General Education course "Understanding Science," instituted to prepare future citizens for democracy in the age of the atomic bomb. Conant, a chemist, president of Harvard, and chief administrator of the Manhattan Project during World War II, was also an innovative historian of chemistry. His case study on the demise of the phlogiston theory in James Bryant Conant, ed., *Harvard Case Studies in the History of the Experimental Sciences* (Cambridge, MA: Harvard University Press, 1950) provided Kuhn with some of his most powerful examples. The attractions of the post–Los Alamos physics for the brightest young minds in the late 1940s and '50s need not be belabored. On the Cold War background to Kuhn's work, see George Reisch's chapter in this volume, and on Kuhn's early career as a physicist, see Peter Galison's chapter in this volume.

11. In addition to Steven Shapin and Simon Schaffer, *Leviathan and the Air Pump* (Princeton: Princeton University Press, 1985), key works included Nancy Cartwright, *How the Laws of Physics Lie* (Oxford: Clarendon Press, 1983); Ian Hacking, *Representing and Intervening: Introductory Topics in the Philosophy of Science* (Cambridge: Cambridge University Press, 1983); Harry M. Collins, *Changing Order: Replication and Induction in Scientific Practice* (London: Sage, 1985); Peter Galison, *How Experiments End* (Chicago: University of Chicago Press, 1987); David Gooding, Trevor Pinch, and Simon Schaffer, eds., *The Uses of Experiment: Studies in the Natural Sciences* (Cambridge: Cambridge University Press, 1993); and Peter Dear, *Discipline and Experience: The Mathematical Way in the Scientific Revolution* (Chicago: University of Chicago Press, 1995).

12. Kuhn, *Structure*, 161–63, 109, 194–95, 169.

13. Joel Isaac draws a suggestive parallel between the prominence of paradigms in *Structure* and Kuhn's involvement in the Harvard practice of teaching by case studies (Joel Isaac, *Working Knowledge: Making the Human Sciences from Parsons to Kuhn* [Cambridge, MA: Harvard University Press, 2012], 206–10).

14. Kuhn, *Structure*, 120.

15. Ibid., 122. Here, Kuhn is paraphrasing reports in Jacques Hadamard's study of the scientific subconscious (122n13).

16. Margaret Masterman, "The Nature of a Paradigm, " in *Criticism and the Growth of Knowledge*, ed. Imré Lakatos and Alan Musgrave, 59–89 (Cambridge: Cambridge University Press, 1970).

17. Kuhn, *Structure*, 174.

18. Ibid., 191.

19. Ludwig Wittgenstein, *Philosophical Investigations*, trans. G. E. M. Anscombe, 3rd ed. (Englewood Cliffs, NJ: Prentice Hall, 1958), §199, p. 81.

20. Michael Polanyi, *Personal Knowledge* (Chicago: University of Chicago Press, 1956). For its use since Polanyi, especially in science studies, see Harry M. Collins, *Tacit and Explicit Knowledge* (Chicago: University of Chicago Press, 2010).

21. On the long history of paradigms in the tradition of rhetoric, see Ian Hacking's chapter in this volume.

22. Herbert Oppel, *KANΩN. Zur Bedeutungsgeschichte des Wortes und seiner lateinischen Entsprechungen* (Regula-Norma) (Leipzig: Dietrich'sche Verlagsbuchhandlung, 1937), 41.

23. Oppel, *KANN*, 17–20, 32, 67. There is, however, at least one significant novelty in the usage of the Latin *regula* in connection with Roman law, in which the word was used by jurists of the first century CE to collect ancient legal decisions into a general precept or proverb, some two hundred of which were appended to the Justinian *Digest* under the rubric *De diversis regulis juris antiqui* (Heinz Ohme, *Kanon ekklesiastikos. Die Bedeutung des altkirchlichen Kanonbegriffs* [Berlin: Walter De Gruyter, 1998], 51–55).

24. Pliny the Elder, *Natural History*, trans. Harris Rackham, Loeb ed., 10 vols. (Cambridge, MA: Harvard University Press, 1952), book 34, ch. 55; vol. 9, 168–69.

25. [Chevalier de Jaucourt], "REGLE, MODELE (*Synon.*)," in *Encyclopédie, ou Dictionnaire raisonné des sciences, des arts et des métiers*, ed. Denis Diderot and Jean d'Alembert. 35 vols. (Lausanne and Berne: Les sociétés typographiques, 1751–80), 28:116–17.

26. Claudius Galen, *De temperamentis libri III*, ed. Georg Helmreich (Leipzig: B. G. Teubner, 1904), 36. Via Galen's reference to Polykleitos, the word and concept of a canonical male body was taken up by Andreas Vesalius and other early modern anatomists (Sachiko Kusukawa, *Picturing the Book of Nature: Image, Text, and Argument in Sixteenth-Century Human Anatomy and Medical Body* [Chicago: University Of Chicago Press, 2012], 213–18).

27. A. A. Markov, *Theory of Algorithms*, trans. Jacques J. Schorr-Kon and PST Staff (Moscow: Academy of Sciences of the U.S.S.R., 1954), published for the National Science Foundation and the Department of Commerce, U.S.A. by the Israel Program for Scientific Translation, 1.

28. On the continued relevance of the paradigm as "communal model" in the biomedical sciences, see Angela Creager's chapter in this volume.

29. Ludwik Fleck, *Enstehung und Entwicklung einer wissenschaftlichen Tatsache. Einführung in die Lehre von Denkstil und Denkkollektiv* (Basel: B. Schwabe, 1935).

SEVEN

Why the Scientific Revolution Wasn't a Scientific Revolution, and Why It Matters

DANIEL GARBER

When Thomas Kuhn was writing *The Structure of Scientific Revolutions*, it was common to refer to the period in the history of science roughly from Copernicus to Newton as the Scientific Revolution, with a capital "S" and a capital "R." One might reasonably assume that his book might have something to do with what happened during that period, that it might give the reader a theoretical structure to understand the transformations in the understanding of nature that happened in the period.

Since Kuhn published *Structure*, however, a lot has changed in the historiography of the history of science. Over the last twenty-five years or so there has been a chorus of those who want to deny that the events in question—the discoveries of Copernicus, Galileo, Descartes, Bacon, Huygens, Newton, and Leibniz—constituted a scientific revolution. And there is just as loud a chorus of voices on the other side, insisting that there really was a scientific revolution. The ambivalence about the notion of a scientific revolution is nicely captured in the oft-quoted opening pages of Steven Shapin's little book about the period: "There was no such thing as the Scientific Revolution, and this is a book about it."[1] It would be tedious to go through the literature and try to summarize their positions.[2] For some on the rejectionist front, the issue is to deny that there was a radical break between the old ideas and the new and to insist on a kind of continuity between them. Or, perhaps, to deny that the study of nature in the period had the kind of unity and coherence implied by calling it all "science." For others it may be an attempt at a postmodern undermining of science itself as an institution and an argument against an Enlightenment picture of intellectual progress. For those on the other side, defending the idea of the Scientific Revolution is part of defending that same Enlightenment picture and defending the very institution of science itself.

There is no question about it: what we might call the long seventeenth century, extending from the mid-sixteenth to the early eighteenth century, was a period remarkable for its fecundity, something that was recognized even by those who lived through it. Figures such as Francis Bacon and René Descartes certainly recognized their own outstanding intellectual qualities and were eager to tell their contemporaries about the remarkable new start that they were providing, a new and deeper understanding of nature that broke sharply with that of past generations. In 1668, the English poet John Dryden wrote:

> Is it not evident, in these last hundred years (when the Study of Philosophy has been the business of all the *Virtuosi* in *Christendome*) that almost a new Nature has been reveal'd to us? that more errours of the School have been detected, more useful Experiments in Philosophy have been made, more Noble Secrets in Opticks, Medicine, Anatomy, Astronomy, discover'd, than in all these credulous and doting Ages from *Aristotle* to us? so true it is that nothing spreads more fast than Science, when rightly and generally cultivated.[3]

The progress represented by scientific investigations of the seventeenth century was a standard trope in the Enlightenment of the eighteenth century.[4] This is, in essence, the master narrative that survived until quite recently.

It is beyond doubt that extraordinary things happened in the seventeenth century. But this is the question: was it a *revolution*?[5] In asking this question, I don't mean to ask just whether it was a revolution in the strict Kuhnian sense, the move from one paradigm, through a crisis, and into another incommensurable paradigm. My question is more basic still: is it illuminating to analogize the intellectual changes that happened in European science during that period to what goes on when one political order violently replaces another?

A political revolution happens when one political authority is replaced by another. In the case of the American Revolution, it was King George III, and after a transition period in which it isn't clear who is the authority, he was replaced by the Continental Congress. It wasn't always clear who exactly was in power in France during the French Revolution, but after a certain point it was clear that it wasn't Louis XVI. Or, in Russia, where the czar abdicated and after a short period of instability was replaced within months by a Bolshevik government. There is often a transition period in which it isn't clear who is in charge, but it can't go on forever. A country can't do without some civil authority for very long. Or, to put it another way, human nature abhors a power vacuum.

Kuhn was quite self-conscious about the relation between political revolutions and what he called scientific revolutions, and he saw strong parallels with what happens in scientific change:

> Political revolutions are inaugurated by a growing sense . . . that existing institutions have ceased adequately to meet the problems posed by an environment that they have in part created. In much the same way, scientific revolutions are inaugurated by a growing sense . . . that an existing paradigm has ceased to function adequately in the exploration of an aspect of nature to which that paradigm itself had previously led the way. In both political and scientific development the sense of malfunction that can lead to crisis is prerequisite to revolution.[6]

In a scientific revolution, as in a political one, the crisis is eventually resolved by the adoption of a new order, a new paradigm in the case of a scientific revolution. And just as in a political revolution the new political order involves fundamental changes in political life and institutions, in a scientific revolution we are dealing with a choice between incommensurable paradigms, Kuhn argued. In a celebrated passage, Kuhn wrote:

> Like the choice between competing political institutions, that between competing paradigms proves to be a choice between incompatible modes of community life.[7]

Just as in a political revolution we have the passage from an old regime, through a crisis to a new regime, in a scientific revolution we pass from an old paradigm, through a crisis to a new paradigm.

But does this pattern fit the so-called Scientific Revolution of the sixteenth and seventeenth centuries?

Now, there are many dimensions to the changes that happened in the period: changes in theoretical structures; changes in scientific practice; changes in the organization of universities, scientific societies, patronage, and laboratories; changes in the relations between different subdisciplines and between the arsenal and workshop and the study; changes in scientific communication, from letter to learned journal; and changes in instrumentation, from the microscope and telescope to the barometer and air pump, among many other changes. But I would like to focus on one change that was of particular interest to the participants in the scientific enterprise in the period: the eclipse of Aristotelian natural philosophy and its challenge by what at least some of the actors in the period called "the new philosophy." I don't in any way mean to suggest that this particular change is somehow definitive of the so-called Scientific Revolution; there were a lot important changes in the period, and there is a continuing lively debate about which were crucial and which less so. But the attack on Aristotle and Aristotelian natural philosophy is something that is quite prominent both for the actors in the period and for later commentators. The question I want to ask is whether it is illuminating to think of the attack on Aristotelianism on analogy with a political revolution.

Attacks on Aristotle and his followers are prominent, for example, in figures like Bacon and Galileo, to take two of the most visible cases. Bacon emphasized the sterility of Aristotelian natural philosophy, the fact that it didn't lead to works. In a memorable passage from the *Instauratio magna* (1620) he wrote:

> I must openly declare that this wisdom, derived mainly from the Greeks, is what might be called the boyhood of science and, as with boys, it is all prattle and no procreation. For productive of controversies, it is barren in works; so that the fable of Scylla seems to catch the present condition of letters to the life for she had the face and countenance of a virgin but a womb begirt with barking monsters.[8]

Bacon promised readers that his method would lead to a new natural philosophy, one that would allow us to control nature. Galileo famously included the ancient Aristotelian commentator Simplicio in his *Dialogue Concerning the Two Chief World Systems* (1632) and his *Dialogue Concerning the Two New Sciences* (1638), in order to mock the Aristotelian professors of his day. Descartes's career suggests even more closely the revolutionary pattern of old paradigm, crisis, and revolution. In his *Discours de la méthode* (1637), Descartes presented himself as the smart student who goes to school and discovers that the emperor has no clothes—scholastic learning is completely empty—then goes off into the world and discovers a new philosophy, one that will replace the philosophy of Aristotle taught in the schools. Later he devoted three years of his life to writing a textbook, his *Principia philosophiae* (1644), which was intended to replace the textbooks used to present Aristotelian philosophy in the schools. In this way Descartes conceived of himself as the new Aristotle, the new authority.[9]

But there is another point of view. Descartes, Bacon, and Galileo were by no means the only challengers to Aristotelian natural philosophy. By the time Descartes was writing, there were many. In 1625, Marin Mersenne, later to become Descartes's close friend and supporter, published *La vérité des sciences*, whose aim was, in part, to defend the very Aristotelian learning that Descartes was to reject in 1637. In the course of that polemic, he made the following comments about the group of contemporary philosophers who have attempted to reject the philosophy of Aristotle:

> Francesco Patrizi tried to discredit this philosophy [i.e. that of Aristotle], but he didn't succeed any more than did Basson, Gorlaeus, Bodin, Charpentier, Hill, Oliva, and many others, who through their quills, only left monuments to the fame of this philosopher, who were not able to fly high enough to dampen the soaring and glory of the Peripatetician, since he transcends everything sensible and imaginable, and the others crawl on the ground like little worms.

Mersenne ended the passage with a famous line: "Aristotle is an eagle in philosophy, and the others are like chicks, who wish to fly before they have wings."[10] While the metaphors are mixed, the point is clear: the lone eagle is not the philosopher, who, like Descartes or Bacon or Galileo, challenges tradition and tries to find an alternative to orthodoxy; the true eagle is Aristotle himself, and those who challenge him are part of a flock of mere wingless chicks.

Mersenne was not alone in seeing those who challenge Aristotelian orthodoxy as belonging to a kind of club. The list of opponents to Aristotelianism that Mersenne gives here is remarkably stable until the 1680s, a collection of heterodox (and heterogeneous) thinkers who are grouped together again and again. (See the appendix at the end of this chapter for a partial catalog of some of the lists that one can find in the period.)

Though he doesn't name names there, Mersenne used the Latin term *novatores*, or "innovators," in his 1623 commentary on Genesis to refer to this group of people when he wrote, "Since many, indeed virtually all of these *novatores* listen badly and have incorrect views about the Catholic faith, they have, indeed, given a hand to the Calvinists, the Lutherans, the Arminians, or to other perfidious heretics."[11] Writing just a couple of years later in his *Advis pour dresser une bibliotheque* (1627), Gabriel Naudé called those on his version of the list "*novateurs*," the French equivalent of "novatores."[12] In English they were often called "Novellists."[13] The terminology varies from language to language and person to person. But what seems important about this group of people who oppose Aristotelianism is that they were doing something new. And it wasn't necessarily considered a good thing.

Bernardino Telesio, whom Bacon listed as "the first of the new philosophers" and "the best of the *Novellists*"[14] is on virtually all of the lists, as are
Francesco Patrizi and Tommaso Campanella. So are Giordano Bruno and
William Gilbert. Among older figures, Girolamo Fracastoro, Petrus Ramus,
and Girolamo Cardano occasionally appear, but interestingly enough, only
very rarely does Paracelsus show up. Many later figures also appear with
notable frequency. Among the better-known figures are Johannes Kepler,
Galileo, and Bacon himself. As his reputation spread, Descartes made the
list, as did occasionally Pierre Gassendi. But there are many lesser-known
figures who appear with great regularity, including Sébastien Basson,
Nicholas Hill, Nathanael Carpentarius, David Gorlaeus, and Godifredus
Chassinus.

It is an interesting and diverse group. One might think of the *novatores*
as a kind of alternative party to the Aristotelians. But there was an important difference. As different as the Aristotelians might have been from one
another, they had texts in common: in natural philosophy the authoritative texts of the *De anima* and the *Physica* that they shared. Among the *novatores*, the only thing that they had in common what that they rejected the
authority of Aristotle and the Aristotelians; beyond that, there was little in
the way of a common thread. Telesio explained everything in terms of hot
and cold, Gilbert explained everything in terms of magnetism. Others, like
Basson, Gorlaeus, and Gassendi, were some variety or another of atomist.
Galileo wasn't really a natural philosopher in the sense of offering a system
of explanation at all but did offer non-Aristotelian doctrines of cosmology and motion. Dona Oliva was a medical writer whose views were anti-
Galenic. Figures like Bacon, Galileo, and Descartes are usually grouped together as a kind of "progressive wing" of the new philosophers, what many
twentieth-century commentators have in mind when they talk about the
New Philosophy. But when we examine them more carefully, we have to
acknowledge that their programs were quite distinct and substantially different from one another. Though all the *novatores* from Telesio to Descartes
and beyond agreed in rejecting Aristotle and Aristotelianism, they could
hardly be said to form a uniform school of thought.

Looking at the period from the perspective of the conflict between the
Aristotelians and the *novatores* gives us a very different view of what happened. Historians of philosophy and science often talk about the debate
between traditional science (Aristotelian, etc.) and *the* new science. This, in
a way, is the narrative that we saw earlier in Descartes. For Descartes, there
really was just one alternative to orthodox Aristotelian natural philosophy:
Cartesian philosophy. But that is just the world according to Descartes (and

his followers). From Descartes's point of view, what happened was a revolution, with Descartes himself as the hero, the eagle, the new authority. He
saw it as a scientific revolution in exactly the Kuhnian sense, and he, Descartes, was the winner. For him, the political analogy that moved Kuhn was
just right: he was the hero of his own revolutionary story.

Bacon took the variety of opinions in natural philosophy more into account. In his early "Cogitata et visa" (1607–8?), he noted that the ancient
Greeks were very good at spinning out alternative natural philosophies,
none of which (including, in his opinion, that of Aristotle) were worthy of
being taken seriously: "Their opinions and theories are like the arguments
of so many stage-plays, devised to give an illusion of reality, with greater
or less elegance, carelessness or coarseness." So it was with the ancients, at
least until Aristotle eliminated all the competition, "like an Ottoman Turk,
in the slaughter of his brethren, and with success." This multiplicity, he
argues, is just the situation with the moderns:

> Here the analogy of Astronomy can help us. One party wants the earth to re-
> volve, another wants to explain the apparent motions by eccentrics and epi-
> cycles, and what is visible in the heavens supports either opinion equally and
> gives no casting vote. Nay, even the calculations based on tables of observa-
> tions suit either view. So it is with Natural Philosophy. Here, and with even
> greater facility, theories can be thought up, all different, all self-consistent,
> all pulling men's minds in different directions, and all appealing for support
> to the vulgar observations which in questions of this sort are allowed the
> role of judge. Have there been in our time, or the preceding generation any
> lack of men to think up new systems of nature?[15]

For Bacon, this was just another sign of how problematic things were in the
modern world: the multiplicity of new philosophies, which he then went
on to catalog in this essay, challenges contemporary readers to resolve the
chaos by finding a way to a unique, new, and correct natural philosophy.
And Bacon was quite willing to be the one to lead his readers to this new
and improved philosophy.

But it didn't look that way to everyone. The very existence of a multiplicity of alternatives to Aristotelian natural philosophy without an obvious way of choosing among them led others to think of knowledge in
a more open-ended way. One can see this new antiauthoritarianism in
Théophraste Renaudot's *conférences* in Paris in the 1630s. In these weekly
meetings, open to anyone with time on his hands, Renaudot offered a few
hours of entertaining conversation on a variety of intellectual topics at his

Bureau d'Adresse at the rue Calandre, "au Grand Coq," on the Ile de la Cité. At each *conférence*, people came prepared to present their different points of view and defend them. As many as seven or eight different views on a single question would be presented and discussed in a single meeting. Starting in 1634, a large group of these *conférences* were published in five volumes. A number of the volumes were translated into English and were popular in England as well.[16]

Renaudot's *conférences* treated a wide variety of subjects, both serious and frivolous, from Copernicanism and the cause of magnetic attraction to whether men or women are more inclined to love. They showed a remarkable openness of spirit and a willingness to entertain different points of view. While the Aristotelian education of many of the participants was in evidence, there were many other points of view as well. Renaudot is explicit about not wanting to limit discussion to the opinions of the schools. In the letter to the reader in the first collection of *conférences* he wrote:

> Certain people might have desired that one not be allowed to advance an opinion contrary to that of the School. But that would seem to be inconsistent with the freedom of our reason, which loses its name if it remains entirely captive under the rod of the authority of a magistrate; the temperament of our nation is less well suited to this than is that of any other nation. And daily experience shows us that there is nothing more inimical to knowledge than to prevent the search for truth, which principally appears in the opposition between contraries.[17]

One speaker expressed what might be the spirit of the *conférences*, a kind of amiable eclecticism. All philosophical sects have their weaknesses, this speaker argued, and so he concluded:

> The best of all is to be none of them, but in imitation of the bees, gather the good from each, without becoming attached to any. . . . This is also the path Aristotle followed in all of his philosophy, and principally in his *Physics* and *Politics*, which are nothing but a collection of the opinions of the ancients. . . . Also, we are no more obligated to embrace the philosophy of Aristotle than he did that of his predecessors, and we are permitted to add one of his precepts to some from Raimond Lull, or Ramus, and all the others.[18]

The actual views of specific *novatores* don't explicitly come up very often in the course of the discussions; actually, specific citations almost never come up in discussion. However, one can see in these discussions an appreciation of the pleasures of novelty and innovation, and a nondogmatic love of debate and the consideration of alternative points of view.

But the *novatores* are the explicit concern of Charles Sorel, an eclectic encyclopedist who wrote explicitly about the *novatores* in the 1650s. Sorel was an extremely prolific writer who first burst into print in 1622 at the age of twenty or maybe a little bit older, with the daring romance entitled *Francion*, which became a great success, as did a number of literary works that followed. But in his mid-thirties, Sorel turned away from the libertine literary life he had led and toward what he considered more mature and serious stuff. In 1634, he began a series of books that collectively made up what he called *La science universelle*, a project that occupied him for more than twenty years.[19] One of the summary essays he wrote at the end of this project was entitled "Le Sommaire des opinions les plus estranges des Nouateurs en Philosophie."[20] In that work he added his own list of *novateurs*: Telesio, Patrizi, Cardano, Ramus, Copernicus, Galileo "& autres Astonomes [Kepler is included in the chapter]"; Bruno, Gorlaeus, Carpenter; "Enchyridion de la physique restituée [Jean d'Espagnet]"; Basson, Campanella, Descartes; "les nouvaters chymistes, de Paracelse & autres, & particularement d'Estienne de Claves, Henry de Rochaz [Rochas]." While it is somewhat larger than some of the earlier lists, it is quite recognizable as the successor of some of the lists we saw earlier.

Sorel had very positive comments about his *novateurs*. He admitted that while the views of some are "fantastiques et imaginaires," "the others address themselves to solid truths, and are to be praised the more for their being hidden."[21] He continued:

> Although the very name of "novateur" might be odious to many people, we must be careful that even if it is to be feared in matters concerning theology, it isn't so in natural and human philosophy.[22]

He admitted that there were some who are *novateurs* simply out of a spirit of contradiction.[23] But he praised others for their courage to point out the errors of Aristotle. While Sorel did not agree with Telesio's natural philosophy, he wrote that "we must praise the grandeur of Telesio's courage, for having dared to be the first to criticize the errors of the ancients."[24] Sorel ended his essay with a plea for being open-minded. One shouldn't accept the ancients dogmatically, nor should we reject them all. "One should take the middle way in this matter," accept a view when it warrants being accepted, and suspend judgment in all things uncertain.[25]

So, what's my point? While there were some who pretended to the status of the new Aristotle, I would claim that the challenge to Aristotelianism never really resolved in favor of a new authority. Descartes was certainly a leading contender for a long while, but no one could reasonably claim that he actually succeeded at his ambition. Newton is often credited with having won and having closed the Scientific Revolution. If we are looking for a winner, he is a plausible candidate, but I have serious doubts about even Newton. The Newtonian program never had the scope that the Aristotelian program had. The Baconian experimental philosophy certainly persisted. And so did chemistry (or "chymistry", as some now want to call it[26]), originally a competing program in natural philosophy that has persisted alongside the mathematical physics in the Newtonian style. And the life sciences, an integral part of the Aristotelian project, were quite distinct from the Newtonian project in the eighteenth century. Even if we look more narrowly at physics throughout the eighteenth century, Leibnizian natural philosophy was a serious challenge to the Newtonian program throughout the eighteenth century, at least on the Continent, and much of what we think of as the Newtonian program may represent a kind of compromise between the two.[27] I would claim that the diversity of alternative anti-Aristotelian programs that blossomed in the late sixteenth and early seventeenth centuries never completely sorted itself out into a single alternative to the Aristotelian program, nothing that could be called *the* new science. And, I would claim, this situation has persisted. What we can see in the seventeenth century is the early history of what has come to be called among contemporary philosophers of science the disunity of science, the idea of the scientific enterprise as a bundle of competing programs with different methodological, theoretical, practical bases—in Kuhnian terms, competing paradigms that never fully resolve.[28]

In this way, the period that is generally called the Scientific Revolution looks less like a real revolution—an old regime that enters into crisis before being replaced by a new regime—and more like the Protestant Reformation that happened at roughly the same time. Luther and Calvin challenged the Roman Church and established churches of their own. But they didn't succeed in replacing the Roman Church with a new and reformed church; indeed, they didn't form a unified opposition to the Catholic Church either: there isn't anything that you can call *the* Protestant Church. There are Lutherans (and varieties of them) and Calvinists (and varieties of them), and there are numerous other Protestant sects as well, all of which coexist with a Roman Catholic Church. In this way, the Reformation fragmented the religious community. And, in a way, one can say that with the *novatores* the intellectual community was fragmented as well: Aristotelians persisted for quite some time, but they were joined by a variety of different sects of philosophers, and the diversity of views endured. Cartesianism was certainly important, just as Lutheranism was. But there were also devotees of Galileo who followed a more mathematical path, or experimentalists and natural historians who looked to Bacon, or chymists and various other kinds of students of nature, just as there were many very different kinds of Protestants. (I would argue that the connection between the Protestant Reformation and the blossoming of new sciences is closer still, but that's another story.)

There is, I think, a lesson for the Kuhnian program here. In politics, a crisis in leadership must be resolved: when one government falls, it must be replaced, and relatively quickly. When it isn't, there is social chaos. But the situation with respect to a Kuhnian paradigm is different: there is no reason why the crisis surrounding a Kuhnian paradigm has to be resolved quickly, or, for that matter, at all. There is no reason why the larger scientific community can't live in suspense for quite some time or why it can't sustain multiple competing programs in some domain, or multiple domains that overlap and compete. That is, there is no reason why Kuhnian revolutions need ever be resolved. From the point of view of the so-called disunity of science movement, this is exactly the way science works, as a collection of incompatible and competing programs that coexist and evolve, without ever reaching the fabled unity that the logical positivists once championed.

Would this have disturbed Kuhn? Not necessarily. The main example that I have been exploring is the Scientific Revolution of the sixteenth and seventeenth centuries. It is very interesting that nowhere in *Structure* does Kuhn ever so much as mention this as an example of what he is talking about. In general, the examples he gives are for somewhat narrower revolutions, like the Copernican, Newtonian, Darwinian, or the Einsteinian revolutions. To that extent it might not have been that important (or surprising) to Kuhn to discover that the so-called Scientific Revolution wasn't a revolution. But in another way, I think that these observations hit fairly deeply at the implicit underlying claim that science changes through revolutions, an old paradigm coming into crisis, and being replaced by a new paradigm. No doubt there are Kuhnian revolutions in science, circumstances in which a new paradigm wins and an old paradigm loses and disappears. When that happens, we have what Kuhn calls normal science, science conducted within a single dominant paradigm. But, at the same time, there is no reason why challenges to the authority of an old paradigm have to end with a clear winner: in the scientific world, a diversity of competing alternatives and not Kuhnian normal science may turn out to be the norm.[29]

APPENDIX

Novatores: A Selection of Lists

Francis Bacon

"Cogitata et visa" (1607–8?)[30]

Telesio, Fracastoro, Cardano, Gilbert

Historia naturalis et experimentalis (1622)[31]

Patrizi, Telesio, Bruno, Severninus the Dane, Gilbert the Englishman, Campanella

Marin Mersenne

Quaestiones celeberrimae in Genesim (1623)[32]

Campanella, Bruno, Telesius, Kepler, Galileo, Gilbert, et al.; Bacon, Fludd, Hill, Basson

L'impiété des déistes, athées, et libertins de ce temps (1624)[33]

Gorlaeus, Charpentier, Basson, Hill, Campanella, Bruno, Vanini, "& quelques autres"

La vérité des sciences (1625)[34]

Patrizi, Basson, Gorlaeus, Bodin, Charpentier, Hill, Oliva

Gabriel Naudé

Apologie pour tous les grand hommes qui ont esté accusez de magie (1625)[35]

Telesio, Patrizi, Campanella, Bacon, Bruno, Basson

Advis pour dresser une bibliothèque (1627)[36]

Telesius, Patrice, Campanella, Verulam, Gilbert, Jordan Brun, Gassand, Basson, Gomesius, Charpentier, Gorlée

Jean-Cécile (Ianus Caecilius) Frey

Cribrum philosophorum qui Aristotelem superiore et hac aetate oppugnarunt (1628)[37]

Campanella, Patrizi, Bacon, Telesio, Ramus, Godefridus Chassinus, "the Vile Villon," de Clave, Gassendi, Pomponazzi, and Valla, Raimon Llull (incorrectly given as "Ludius"), Basson

René Descartes

Letter to Beeckman, October 17, 1630[38]

Telesius, Campanella, Brunus, Basso, Vaninus, Novatores omnes

ÉTIENNE DE CLAVE

Paradoxes ou Traittez Philosophiques des Pierres et Pierreries, contre l'opinion vulgaire (1635)[39]

Patrizi, Basso, Campanella, Gassendi, Dona Catharina Oliva, "personages Crestiens & Philosophes . . ."

JEAN BACHOU

Preface to the French edition of Jean d'Espagnet, *La Philosophie naturelle restablie en sa pureté* (1651)[40]

Telesio, Patrizi, Campanella, Bacon, Fludd, Gorlaeus, Taurellus, Ramus, Descartes, Sorel, Jean d'Espagnet

GUY HOLLAND

The grand prerogative of humane nature namely, the souls naturall or native immortality (1653)[41]

Telesio, Patrizi, Ramos, Basson, Gassendi, Descartes, Regius, Campanella

ADRIEN HEEREBOORD

"Consilium de ratione studendi philosophiae" (1654)[42]

Vives, Ramus, Patrizi (in his *Discussiones Peripateticae*), Gorlaeus, Campanella, Telesio, Basson, and the brothers Boot are *novatores* who are just destructive of the old; Patrizi (in his *Nova Philosophia*), Bacon, Comenius, and Descartes additionally set out something new.

JOHN WEBSTER

Academiarum examen, or the Examination of the Academies (1654)[43]

Patrizi, Ficino, Descartes, Regius, Phocylides Holwarda, Gassendi, Telesio, Campanella, Gilbert, Paracelsus

CHARLES SOREL

"Le Sommaire des opinions les plus estranges des Nouateurs en Philosophie" (1655)[44]

Telesio, Patrizi, Cardano, Ramus, Copernicus, Galileo "& autres Astonomes [including Kepler]"; Bruno, Gorlaeus, Carpenter, "Enchyridion de la physique restituée [Jean d'Espagnet]"; Basson, Campanella, Descartes, "les nouvateurs chymistes, de Paracelse & autres, & particulierement d'Estienne de Claves"; Henry de Rochaz [Rochas], Villon

Notes

I would like to thank the audience at the University of Chicago Kuhn Conference for very helpful discussion, and the anonymous reviewer for the University of Chicago Press. I would especially like to thank Roger Ariew, who commented on an earlier draft of this essay. An earlier and much cruder version of some of the material in this paper was presented in "Galileo, Newton and All That: If It Wasn't a Scientific Revolution, What Was It? (A Manifesto)," *Circumscribere* 7 (2009): 9–18.

1. Steven Shapin, *The Scientific Revolution* (Chicago: University of Chicago Press, 1996), 1.

2. For two summary accounts of some of the main positions on this question, see the introductory essays, Steven J. Harris, "Introduction: Thinking Locally, Acting Globally", *Configurations* 6 (1998): 131–39, and Donald A. Yerxa, "Introduction: Historical Coherence, Complexity, and the Scientific Revolution", *European Review* 15 (2007): 439–44, and the collections of papers that follow each.

3. John Dryden, *Of Dramatick Poesie, an Essay* (London: Henry Herringman, 1668), 9, quoted in H. Floris Cohen, *The Scientific Revolution: a Historiographical Inquiry* (Chicago: University of Chicago Press, 1994), 1 (original emphasis).

4. See, e.g., the extended discussion of the great scientific figures of the seventeenth century in Jean d'Alembert's *Discours préliminaire* to the *Encyclopédie, ou Dictionnaire raisonné des sciences, des arts et des métiers* (Paris, 1751–72), 1:xxiv ff.

5. I am also sympathetic to the idea that one should not think of what is being transformed in the period as "science" in anything like the modern sense, if there is such a thing. While there were a variety of knowledge practices relating to the natural world—natural philosophy, mixed mathematics, natural history, medicine, and so forth—it is wrong to think of them as if they constituted a unified and coherent domain. Which is to say, what happened in the sixteenth and seventeenth centuries may have been *neither* a revolution *nor* scientific, strictly speaking.

6. Thomas Kuhn, *The Structure of Scientific Revolutions*, introduction by Ian Hacking, 4th ed. (Chicago: University of Chicago Press, 2012), 92.

7. Ibid., 94.

8. Francis Bacon, *The Oxford Francis Bacon*, ed. Graham Rees with Maria Wakeley, vol. 11, *The* Instauratio magna *Part II* : Novum organum *and Associated Texts* (Oxford: Oxford University Press, 2004), 11–13.

9. See, e.g., Daniel Garber, *Descartes' Metaphysical Physics* (Chicago: University of Chicago Press, 1992), ch. 1.

10. Marin Mersenne, *La vérité des sciences* (Paris: Toussainct du Bray, 1625), 109–10, cf. Galileo's comparison of the philosopher with an eagle in the *Assayer* (1623), in Stillman Drake, *Discoveries and Opinions of Galileo* (Garden City, NY: Doubleday, 1957), 239.

11. Marin Mersenne, *Quaestiones celeberrimae in Genesim* (Paris: Sebastian Cramoisy, 1623), "Ad lectorem," unpaginated.

12. Gabriel Naudé, *Advis pour dresser une bibliotheque* (Paris: François Targa, 1627), 135.

13. Francis Bacon, *Sylva Sylvarum or a Natural History in Ten Centuries* (London: W. Lee, 1626), expt. 69, calls Telesio a "novellist."

14. See Bacon, "De principiis et originibus," in *The Oxford Francis Bacon*, vol. 6, *Philosophical Studies c. 1611–c.1619* (Oxford: Oxford University Press, 1996), 258–59; Bacon, *Sylva Sylvarum*, expt. 69.

15. Bacon, *The Works of Francis Bacon*, ed. J. Spedding, R. L. Ellis, and D. D. Heath (London: Longman and Co. et al., 1858–74), 3:602–3, trans. in Benjamin Farrington,

The Philosophy of Francis Bacon: An Essay on Its Development from 1603–1609 (Liverpool: Liverpool University Press, 1964), 84, 85.

16. For further details on Renaudot and his *conférences*, see Howard M. Solomon, *Public Welfare, Science, and Propaganda in Seventeenth Century France: the Innovations of Théophraste Renaudot* (Princeton, NJ: Princeton University Press, 1972); Simone Mazauric, *Savoirs et philosophie à Paris dans la première moitié du XVIIe siècle: les conférences du Bureau d'adresse de Théophraste Renaudot, 1633–1642* (Paris: Publications de la Sorbonne, 1997); Kathleen A. Wellman, *Making Science Social: The Conferences of Théophraste Renaudot, 1633–1642* (Norman: University of Oklahoma Press, 2003); and my review of this last item, Daniel Garber, "Review of Wellman, *Making Science Social*," *Early Science and Medicine* 10 (2005): 428–34.

17. Théophraste Renaudot, *Premiere centurie des questions traitees ez conferences du Bureau d'Adresse, depuis le 22. iour d'Aoust 1633 iusques au dernier Iullet 1634* (Paris: Au Bureau d'Adresse, 1635), "Avis au Lecteur," unpaginated.

18. Théophraste Renaudot, *Seconde centurie des questions . . .* (Paris : Au Bureau d'Adresse, 1636), 313.

19. For a brief account of Sorel and his career, see Gabrielle Verdier, *Charles Sorel* (Boston: Twayne, 1984). For an account of his *Science universelle*, see Mariassunta Picardi, *Le libertà del sapere: filosofia e "scienza universale" in Charles Sorel*. (Naples: Liguori, 2007).

20. In Charles Sorel, *De la perfection de l'homme* (Paris: Robert de Nain, 1655), 209–75. The essay was reprinted with some interesting changes thirteen years later in Sorel, *La science universelle, tome quatriesme* (Paris: Theodore Girard, 1668), 360–449.

21. Sorel, *De la perfection*, 210; cf. 267.

22. Ibid., 210.

23. Ibid., 267, cf. 210.

24. Ibid., 267; cf. 218.

25. Ibid., 273–4.

26. William Newman and Lawrence Principe, "Alchemy vs. Chemistry: The Etymological Origins of a Historiographic Mistake," *Early Science and Medicine* 3 (1998): 32–65.

27. I am basing my claim here on some very exciting work in progress, including especially work by Marius Stan and Anne-Lise Rey, which shows the continuing debates between Newtonian and Leibnizian strains of natural philosophy throughout the eighteenth century.

28. On this strain of thought in recent philosophy of science, see especially John Dupré, *The Disorder of Things: Metaphysical Foundations of the Disunity of Science* (Cambridge, MA: Harvard University Press, 1993); and Peter Galison and David Stump, eds., *The Disunity of Science: Boundaries, Contexts, and Power* (Stanford, CA: Stanford University Press, 1996).

29. This resonates in an interesting way with Feyerabend's idea that because the world we see is in part shaped by the theories we hold, a good empiricist should want to encourage as many alternative theoretical perspectives as possible, however wacky they may seem. See Paul Feyerabend, *Against Method* (London: Verso, 1975), ch. 3. Feyerabend's comment is intended to be normative, though, while mine is an observation about what I take to be the actual situation in the investigation of nature.

30. In Bacon, *The Works of Francis Bacon*, 3:603.

31. In Bacon, *The Oxford Francis Bacon*, vol. 12, *The* Instauratio Magna *Part III:* Historia

naturalis et experimentalis, Historia ventorum, Historia vitae & mortis (Oxford: Oxford University Press, 2008), 9.

32. Mersenne, *Quaestiones celeberrimae in Genesim.* The first six names appear in the unpaginated "Praefatio et prolegomena ad lectorem"; the last four names appear in col. 1838.

33. Marin Mersenne, *L'impiété des déists, athées, et libertins de ce temps* (Paris: Pierre Billaine, 1624), 237–38.

34. Mersenne, *La vérité des sciences,* 109.

35. Gabriel Naudé, *Apologie pour tous les grand hommes qui ont esté accusez de magie* (Paris: François Targa, 1625), 331.

36. Naudé, *Advis pour dresser une bibliotheque* (Paris: François Targa, 1627), 135.

37. Jean-Cecile Frey, *Cribrum philosophorum qui Aristotelem superiore et hac aetate oppugnarunt,* in *Opuscula varia nusquam edita* (Paris: Petrus David, 1646), 29–89. The *Cribrum* is a set of lectures given by Frey in Paris in 1628, but published after his death by his students, presumably based in their lecture notes.

38. René Descartes, « Letter to Beeckman, 17 Oct. 1630," in *Oeuvres de Descartes,* 1:158.

39. Étienne de Clave, *Paradoxes ou Traittez Philosophiques des Pierres et Pierreries, contre l'opinion vulgaire* (Paris: Pierre Chevaliere, 1635), 186.

40. Jean Bachou, Preface to the French edition of Jean d'Espagnet, *La Philosophie naturelle restablie en sa pureté* (Paris: Edme Pepingué, 1651), "Discours a la recommendation de la Philosophie ancienne restablie en sa pureté; Et sur le nom de son premier Autheur," unpaginated. Bachou is the French translator of the work, originally published anonymously in Latin; the "Discours" is his introduction.

41. Guy Holland, *The grand prerogative of humane nature namely, the souls naturall or native immortality* (London: Roger Daniel, 1653), 89.

42. Adrien Heereboord, "Consilium de ratione studendi philosophiae," in *Meletemata philosophica* (Leiden: Franciscus Moyardus, 1654), 1:28.

43. John Webster, *Academiarum examen, or the Examination of the Academies* (London: Giles Calvert, 1654), 106.

44. Charles Sorel, "Le Sommaire des opinions les plus estranges des Nouateurs en Philosophie," in *De la perfection,* 209–75.

Kuhn on the UC Berkeley campus with Sproul Plaza in the background.

EIGHT

Paradigms and Exemplars Meet Biomedicine

ANGELA N. H. CREAGER

In "The Route to Normal Science," the second section of *The Structure of Scientific Revolutions*, Kuhn said that by choosing the word "paradigm," he meant to suggest "that some accepted examples of actual scientific practice—examples which include law, theory, application, and instrumentation together—provide models from which spring particular coherent traditions of scientific research."[1] His subsequent discussion of paradigms emphasized both their role in training students to join specific research communities and their function in guiding scientific expectations and theories. A paradigm in this sense constituted a model or set of models by which to solve problems. In his postscript to the second edition of the book, Kuhn revisited his notion of paradigms, conceding that he used the word to refer both to large-scale encompassing conceptual frameworks and to local points of reference (models). He suggested that "exemplar" might be a better term for the latter, designating those shared examples that bind together and orient research communities. This, he emphasized, may be the most important insight of his much-discussed book: "A paradigm as shared example is the central element of what I now take to be the most novel and least understood aspect of this book. Exemplars will therefore require more attention than the other sorts of components of the disciplinary matrix."[2] This aspect of Kuhn's conception of scientific change, the exemplar as a kind of communal model, has particular relevance to biology and medicine.

Several commentators—for our purposes, simply take Ian Hacking's snappy introduction to the fiftieth-anniversary edition of *Structure*—note that Kuhn drew almost exclusively on the physical sciences in illustrating his theory of scientific change.[3] Kuhn himself admitted that he did not include evidence from biology, "partly to increase this essay's coherence and partly on grounds of present competence."[4] However, his elaboration on paradigms in the 1969 postscript made telling use of biological considerations. In commenting on how scientific communities exist at numerous scales, from all natural scientists to disciplines to subgroups such as protein chemists, solid-state physicists, and radio astronomers, he stated:

> It is only at the next lower level that empirical problems emerge. How, to take a contemporary example, would one have isolated the *phage group* prior to its public acclaim? For this purpose one must have recourse to attendance at special conferences, to the distribution of draft manuscripts or galley proofs prior to publication, and above all to formal and informal communication networks including those discovered in correspondence and in the linkages among citations.[5]

It is not surprising that the phage group was a conspicuous example of a research community for Kuhn; Max Delbrück, Salvador Luria, and Alfred Hershey shared the 1968 Nobel Prize in physiology and medicine for their work on bacteriophage. Notably, these three were regarded as the cofounders, even "fathers," of the community of phage biologists who gathered during summer courses and meetings at Cold Spring Harbor Laboratory. What united these researchers was not a conceptual schema or textbook (though James Watson's splendid *Molecular Biology of the Gene*, first published in 1965, did fill that requirement), but their common commitment to a particular experimental subject, namely the T-even phages of *E. coli*. The phage group maintained informal communication of unpublished results through a newsletter and used their gatherings at Cold Spring Harbor and camping trips from Caltech into the mountains of Southern California to communicate findings and draw in new acolytes. In their sharing of materials and information, they consciously emulated the network of *Drosophila* researchers for which Caltech was also prominent.[6]

Phages, *Drosophila*—biologists commonly refer to such well-studied, common research organisms or objects as model systems. These model systems, like strains of laboratory mice, have been standardized through breeding and may be obtained commercially. An important criterion for selection is accessibility, which may depend on wide availability, physical features (size, simplicity, reproduction rate), or simply an acquired prominence within a field of study. Model systems are valued for their particularity, materiality, and possessing a tractable level of complexity. But their key characteristic is surely typicality. Moreover, model systems exhibit a self-reinforcing quality: the more it is studied, and the more perspectives from which it is understood, the more it becomes a model system.[7]

Model systems, I argue, function quite clearly as exemplars in Kuhn's sense, if not exactly in his terms. The symbolic generalizations that characterize physics are rare in biomedical research.[8] Yet Kuhn himself stressed how similarities are embodied in physical situations rather than rules or laws. In the passage on exemplars in his postscript to *Structure*, Kuhn offered up an analysis of how Newton's second law of motion could be extended, by analogy, to apply to new situations, such as free fall, harmonic oscillators, or the gyroscope. Such an extension requires the physics student to learn to "identify forces, masses, and accelerations in a variety of physical situations not previously encountered," as well as to "design the appropriate version of $f=ma$ through which to interrelate them." It involves, in other words, learning to "see a problem as like a problem already encountered."[9] Kuhn then presented a brief historical example, involving Christiaan Huyghens's analysis of Galileo's point pendula and Bernoulli's extension of Huyghens's pendulum to describe the flow of water. "That example," Kuhn asserted, "should begin to make clear what I mean by learning from problems to see situations as like each other, as subjects for the application of the same scientific law or law-sketch. Simultaneously, it should show why I refer to the consequential knowledge of nature acquired while learning the similarity relationship and thereafter embodied in a way of viewing physical situations rather than in rules or laws."[10]

Kuhn's depiction of how scientists extend knowledge to unexplored areas through analogies to well-worked ones describes surprisingly well how model systems function in biomedicine. Take the phage group already mentioned. Some of its members sought to apply to animal viruses the visual method they used to detect and quantify bacteriophage, namely the appearance of phage "plaques" on a petri dish. Renato Delbecco developed the first such animal virus assay, for Western equine encephalitis virus on cultured chicken cells in 1952.[11] The virus caused plaques to develop due to necrosis in the monolayer of cells in tissue culture, similar to the phage infected and lysed *E. coli* cells grown on agar media. As in the bacterial phage assay, each plaque of dead chicken embryo cells represented one "infective unit" of Western equine encephalitis virus[12] (see figure 8.1). Similar tests for other animal viruses emerged, most notably one developed by Dulbecco and Marguerite Vogt for poliovirus. Such plaque assays enabled virologists to determine the titer, or the number of animal viruses, in a given sample. Others invented analogous focus-forming assays for tumor

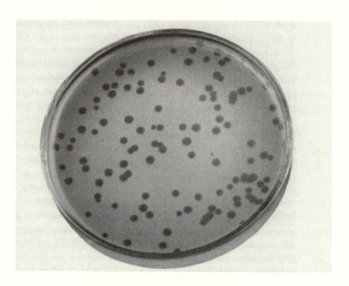

Figure 8.1 (A). Phage plaques on a lawn of *E. coli* cells grown on media in petri dish.
(Gunther S. Stent, *Molecular Biology of Bacterial Viruses*
[San Francisco: W. H. Freeman, 1963], 41.)

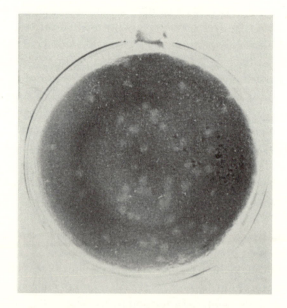

Figure 8.1 (B). Plaques of Western equine encephalitis virus on chicken fibroblasts.
The plaques appear as round clear areas. (Renato Dulbecco, "Production of Plaques in
Monolayer Tissue Cultures by Single Particles of an Animal Virus," *Proceedings of the National
Academy of Sciences U S A* 38 [1952]: 747–52, on 748.)

viruses; the overgrowth of virus-transformed cells, like the plaques of other assays, represented a single infective unit, or virus.[13] As Howard Temin described the assay he and Harry Rubin developed for Rous sarcoma virus: "The foci were the cell culture analogs of tumors in chickens."[14] Moreover, because some bacteriophages could insert themselves into the DNA of host *E. coli* (becoming "latent" until conditions changed), the analogy between bacteriophage and tumor viruses opened up a way to account for how the latter might embed themselves in the host genome—and several tumor viruses and cellular homologues came to be found there.[15]

Bacteriophage was not, however, the only available exemplar for investigations of animal viruses. Frederick Schaffer and Carlton Schwerdt in Wendell Stanley's Virus Laboratory in Berkeley began investigating poliovirus along the lines of previous work with tobacco mosaic virus, treating it as a macromolecule to be analyzed chemically and structurally. This led them to crystallize poliovirus (from surplus vaccine fluid) in 1955, twenty years after Stanley obtained crystals of tobacco mosaic virus[16] (see figure 8.2). In turn, Rosalind Franklin at Birkbeck College, London, obtained poliovirus crystals from Stanley's laboratory early in 1957 and was attempting to use X-ray diffraction to determine the symmetry and structure of poliovirus until her untimely death in April 1958. Her group continued the investigation, showing that poliovirus possesses icosahedral symmetry like that of the spherical tobacco yellow mosaic virus (but not the rod-shaped tobacco mosaic virus).[17] Here biophysical research on plant viruses guided investigations of a human pathogen, opening up structural determinations of poliovirus at the molecular level, whereas a quantitative approach to bacteriophage growth had led other virologists to analyze—and visualize—the agent of polio as a genetic unit of infection.

As can be seen from these examples in virus research, whereas Kuhn largely emphasized the role of theory and problem-solving in terms of conceptual work, a model systems approach brings out the centrality of experimentation at the leading edge of scientific change. What scientists take from exemplars such as bacteriophage or tobacco mosaic virus is not necessarily a way to *see* the world differently as a way to *handle* the world differently—to rearrange it. And even then, such exemplars are not static. Rather than seeing exemplars as fixed or rigid, they are better viewed as changeable points of reference amidst the day-to-day decisions that are constitutive of research. Through such analogies we can discern how model systems shape the direction of research across fields of investigation.

My emphasis on experimentation in biomedicine as more decisive than the testing of theories draws on a historiography that gained traction quite

Figure 8.2 (A). Photomicrograph of Stanley's crystalline tobacco mosaic virus protein, magnified 675 times. (Courtesy of the Bancroft Library, University of California, Berkeley.)

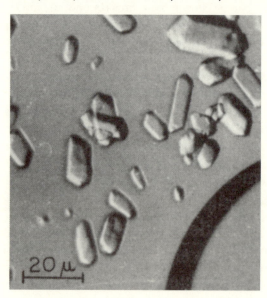

Figure 8.2 (B). Crystals of MEF-1 poliomyelitis virus particles photomicrographed in visible light. (From Frederick L. Schaffer and Carlton E. Schwerdt, "Crystallization of Purified MEF-1 Poliomyelitis Virus Particles," *Proceedings of the National Academy of Sciences U S A* 41 [1955]: 1020–23, on 1021.)

some time after Kuhn's book, especially during the 1990s and 2000s. Peter Galison's *Image and Logic* remains a benchmark for what a historiography of materials looks like, and, not inconsequentially, how it revises Kuhn's depiction of scientific change. In the life sciences, both Hans-Jörg Rheinberger and Robert Kohler drew attention to how experimental systems, as biologists call them, function as local units of research.[18] In one respect, this approach drew on the post-Kuhnian literature on constructivism, insofar as such systems are assembled; they are *material constructs* rather than natural objects.[19] In cases where experimental systems rely on genetically standardized organisms or biochemically purified fractions, the research subjects themselves do not exist outside the laboratory. Second, experimental systems unfold in time and in unpredictable ways.[20] Consequently, scholars have emphasized the element of surprise inherent in biological experimental systems.[21] This provides a different way of viewing what Kuhn called "anomalies"—and seeing them not as catalysts of crisis but rather as motors of scientific change.[22] Shifts in knowledge from such unanticipated results may not be as sweeping as suggested by "paradigm change," but can be transformative nonetheless.

In the case of genetics, Kohler argued that *Drosophila melanogaster* (the fruit fly) colonized the laboratory of T. H. Morgan and displaced other creatures. The high rate of reproduction fueled this transition—the fly stocks became a "breeder reactor" that churned out mutants faster than Morgan's boys could map them.[23] In order to stay on top of the game, these researchers developed patterns of sharing materials and information as well as shipping fly strains to other interested researchers and teachers. In this respect, the social dynamics of experimental genetics derived from the fecund nature of the discipline's premier research subject.[24] At the same time, the limitations of *Drosophila*, once reconstructed as a genetic instrument, led some researchers to move on to other organisms. For example, George Beadle, after struggling with Boris Ephrussi to deploy the fly for biochemical genetics, began working with *Neurospora crassa* (this time with collaborator Edward Tatum), because this bread mold proved more productive of biochemical knowledge.[25]

This account of the emergence of a classic model organism largely shifts the engine of scientific knowledge from the ingenious researcher to the productive material—fly, mouse, mold. Rhetorically, Kohler's argument is carried in part by the fact that flies are organisms: they are mobile, they have a natural history, they have actual biological agency. Thus there is congruence between the figurative and the literal senses in which *Drosophila* took over the lab. Similarly, Karen Rader's account of the standardization of the laboratory mouse, Bonnie Claus's study of the Wistar rat, and Rachel Ankeny's and Soraya de Chadarevian's histories of the nematode worm are animated by actual reproducing organisms.[26] There are equally important differences emphasized by the historical accounts of these experimental organisms. For example, by the 1940s, standardized laboratory mice were produced and sold by the Jackson Laboratory in the tens of thousands, yet scientists used them for a wide range of purposes, unlike in *Drosophila* research—more as pure chemical reagents than as scientific instruments.[27] Even so, one can see key transitions in the history of biology and biomedicine by focusing on the scientific stabilization of these creatures both within and beyond the laboratories that "domesticated" them for research.[28]

There are also diverse experimental systems in biology that cannot be conflated with or signified by a living organism. Rheinberger's history of research on protein synthesis, recounting Paul Zamecnik's investigations of rat hepatic cells, offers just such a case. Rheinberger sought to distinguish the experimental system as a highly local configuration of instruments, assays, and materials, from the organism or material from which a system is constructed (be it rat liver, *E. coli*, or *Drosophila*). In addition, he referred to the ostensible object of study as the *epistemic thing*, which depends on, but is not the same as, the system through which it is being investigated. Epistemic objects are not static, nor do they exist outside the experimental situation. In the case he examined, the epistemic thing was soluble RNA, eventually differentiated and reconceptualized into several molecular entities in the early 1960s, most notably transfer RNA. The experimental system in Zamecnik's laboratory encompassed the materials and practices through which the RNA was visualized: the rat liver tissue, ultracentrifuges, radioisotopes, scintillation counters, and other items and methods employed in the experimentation.[29]

Particularly important to Rheinberger's vision is the inherent *instability* of an experimental system, the constant displacements that actually make it useful for research. Indeed, as soon as "nature" is stabilized sufficiently that an epistemic object is understood, it has ceased to be scientifically interesting. At this point, the epistemic object may become a technical object, used in the experimental setup to search out new epistemic objects. Rheinberger suggested that standardization itself provides limited explanatory power for the understanding how new scientific knowledge emerges from biological experimentation. In this respect his work provides an important counterpoint to the emphasis on metrology in most of the rest of the literature on model organisms and experimental materials in biology.[30]

In different ways, both Kohler and Rheinberger presented the experimental system as a kind of *machine* for research: once established, the system exhibits a self-generating nature. Rheinberger quoted François Jacob's memorable line that experimental systems are "machines for making the future."[31] Yet the "next step" in a trajectory of research is not necessarily dictated by the experimental system. Scientists do not work with their systems in isolation, either from the institutions that support them or, even more importantly, from their colleagues and competitors. Many of the innovations in experimentation result when the researcher borrows strategies, methods, or concepts from someone else's experimental system.

It is in this sense that Kuhn's emphasis on the scientific community and the existence of exemplars proves instructive. The notion of a "model system" can be used to highlight how scientists reconfigure their local experimental systems to mimic the successes achieved by others. To take a classic example from molecular biology, Monod and Jacob's 1961 lac operon model inspired a generation of microbiologists to seek other examples of the regulation of gene expression through repressors and inducers. This both enabled and constrained research on metabolic regulation. Because Monod and Jacob's picture relied fundamentally on the negative regulation (repression), researchers for awhile missed the pervasiveness of positive regulation.[32]

A focus on these kinds of concrete analogies links the attention to experimental systems as units of research to larger scientific changes and transformations. "Experimental systems" bring into view a variegated middle ground that might be obscured under Kuhn's monolithic rubric of "paradigm." When one follows a single trajectory of research, one may subsume even the scientists' actions into the "experimental system." However, when looking at experimental systems as interacting elements in a broader story, one must be able to speak to interactions between scientists who are constantly modifying and recalibrating their experimental systems with reference to the results of others.[33] Rheinberger discussed the mutual reshaping of experimental systems in terms of their "conjunctures." Such systems become "defined with respect to, and in distinction from, other neighboring experimental systems."[34] One way that agreement between laboratories is achieved is that experimental systems are modeled upon each other, facilitating observations of similar biological phenomena in many different species, cells, pathways, and genes. Thus, results "travel" and are generalized through these kinds of material convergences.[35]

The boundaries of an experimental system may be hard to specify. Are human actions inclusive or exclusive? Do experimental systems stop at the laboratory door? Relatedly, the concept of a model system may too easily conflate the research object (fly, virus, gene) with the research apparatus required to study it or with the epistemic object being pursued.[36] Indeed, this kind of slippage is ubiquitous when biologists discuss experimental systems and model systems.[37] This ambiguity, however, points to an important issue in the use of such systems as referents. While model systems remain deeply rooted in their material instantiations, they are, in part, idealizations. To refer to one's system as *Drosophila* is not to call up one specific fly. Neither is the referent entirely abstract. Standing behind the "fly" are many actual flies, and likely many different strains of flies, in conjunction with the relevant practices of genetic experimentation. As Kohler put it, "Genetic instruments are extended systems, consisting of large and fecund families of mutants designed for particular purposes and freighted with ever-growing bodies of craft knowledge and skills."[38] Thus, when researchers refer to their organism as a model system, the local experimental system hovers just in the background; a set of practices is implicit.[39]

The accumulation of practice and knowledge that becomes embedded in the use of particular organisms gives model systems their self-reinforcing quality. As Joshua Lederberg observed in accounting for the postwar popularity of *E. coli*, "Soon, the very accumulation of knowledge, mostly concentrated on a single strain, 'K-12,' made it more likely that it would be a prototype for still further studies."[40] Two important issues should be noted with reference to model system as "prototype." The more a model system is understood, the more useful it becomes as a tool for asking detailed and complex questions. Also, model systems are often treated as "typical" in terms of the vital process they exemplify or the disease they are used to study.[41]

If the value of model systems in biomedicine draws on the practices and know-how that have accrued with their use, it is enhanced by the analogies drawn to humans and their health problems. In the world of postwar medical science, the way in which a researcher's system has been made into a *model* for a particular disease or process is key to its place in research. The way in which a model system mimics or represents human disease often provides the justification linking the government-sponsored mission of improving health with the day-to-day activities of publicly funded scientists. Prior to the pronounced growth of federal government support for biomedical research in the 1950s, public philanthropies such as the National Foundation for Infantile Paralysis and the American Cancer Society began funding research on laboratory models for human pathogens or disease processes. As the National Institutes of Health expanded in both number and budget along a disease-oriented or "categorical" framework, the agency also favored the use of such models. This justification for funding basic research remains central to the National Institutes of Health.[42]

That said, the manner in which a model system's medical relevance is established may not be straightforward. Interestingly, Wendell Stanley himself used the term "model system" when justifying research on well-studied viruses to understand the genesis of cancer. In this case, the analogy between the laboratory situation and the target disease was particularly fraught.[43] Human cancer, after all, fails to exhibit the patterns of contagion or epidemics typical of other viral diseases. Moreover, cancer is not a single disease but a cluster of diseases having to do with unregulated cell growth. Thus, one etiology would be unlikely to explain all incidents of human cancer, a fact not lost on medical practitioners and clinical researchers who questioned viral explanations of cancer.[44] Nonetheless, before Congress in 1959, Stanley strongly advocated the importance to cancer research of models based on tumor viruses and urged the support of such model research by the National Cancer Institute.[45] As he explained a few years later in a lecture for the American Cancer Society:

> I continue to be especially interested in model systems in which the conversion from the normal cell to a cancer cell by means of a virus can be studied in the laboratory in great detail under exactly controlled conditions. Two systems which are especially favorable are concerned with the polyoma virus and with the Rous sarcoma virus. . . . I believe that work on model systems such as those of polyoma and the Rous sarcoma virus may provide ideas useful in studies on human cancer-virus relationships.[46]

Tumor viruses did in fact become a target of intensive research in the 1960s and 1970s, although their significance as a source of human cancer remained contested. The National Cancer Institute launched a Special Cancer Virus Program aimed at developing human cancer vaccines; it ended in failure, however.[47] Nonetheless, model systems cultivated in these efforts, such as SV40 and adenovirus, became principal tools in molecular biology for understanding eukaryotic gene regulation.[48] Tumor viruses were also at the center of debates over the possible public health dangers associated with genetic engineering; these debates resulted in calls for a voluntary moratorium on recombinant DNA experiments with cancer-related animal viruses.[49] Moreover, research in the 1970s into retroviruses in the name of understanding cancer provided crucial referents for understanding—and ultimately treating—HIV/AIDS. As this example shows, histories of model systems can elucidate nonobvious connections among laboratory practice, changing disease etiology, and science policy.

In closing, let me return to Kuhn's observation that key theories or concepts are often embodied in physical situations, such as the pendulum. These exemplars are then extended, by analogy, to new situations. In the same way, model systems function for biologists as exemplars whose utility derive from their typicality, solvability, and unpredictability.[50] My Kuhnian interpretation of model systems in the life sciences does not necessarily answer criticisms of his work. In particular, my use of "model system" has been somewhat imprecise. Is it the organism, the representation, the package of tools and materials, or all of the above? The term and concept share the weaknesses of Kuhn's "paradigm," which is also maddeningly malleable. At another level, I have departed from the spirit of *Structure* in paying so little attention to pedagogy.[51] Not that the role of bacteriophage or the lac operon in teaching molecular biology, for instance, is unimportant. Yet my interest here has been less in the codification of exemplars for pedagogy than in getting at the dynamics of their use by researchers, for which these points of reference, as standards and models, are in flux. It is this understanding of exemplars from Kuhn's *The Structure of Scientific Revolutions* that I find so illuminating for biomedicine.

Notes

This essay as published benefited from comments by Ken Alder, James Chandler, Henry Cowles, Lorraine Daston, Christian Flow, Dave Kaiser, Bob Richards, Bill Wimsatt, and other participants of the Kuhn Conference at University of Chicago on November 30–December 1, 2012. My perspective on model systems and exemplars has been shaped through countless conversations over the years with Jean-Paul Gaudillière, Hans-Jörg Rheinberger, Elizabeth Lunbeck, Mary Morgan, and above all, Norton Wise. This chapter includes material from my book, *The Life of a Virus: Tobacco Mosaic Virus as an Experimental Model, 1930–1965* (Chicago: University of Chicago Press, 2002). It is reprinted here with permission.

1.   Thomas S. Kuhn, *The Structure of Scientific Revolutions*, introduction by Ian Hacking.4th ed. (University of Chicago Press, 2012), 11.

2.   Ibid., 186–87.

3.   Ian Hacking, introduction to Kuhn, *Structure*, ix.

4.   Kuhn, preface to Kuhn, *Structure*, xliii.

5.   Kuhn, *Structure*, 177 (emphasis added). A few years after Kuhn's observation, just the kind of study he had in mind appeared: Nicholas C. Mullins, "The Development of a Scientific Specialty: The Phage Group and the Origins of Molecular Biology," *Minerva* 10 (1972): 51–82.

6.   John Cairns, Gunther S. Stent, and James D. Watson, eds., *Phage and the Origins of Molecular Biology* (Cold Spring Harbor, NY: Cold Spring Harbor Laboratory Press,

1966); James D. Watson, *Molecular Biology of the Gene* (New York: W. A. Benjamin, 1965).

7. Here I am drawing on the brief for a two-year workshop series in Princeton that resulted in Angela N. H. Creager, Elizabeth Lunbeck, and M. Norton Wise, eds., *Science without Laws: Model Systems, Cases, Exemplary Narratives* (Durham, NC: Duke University Press, 2007).

8. Some scholars would argue that they are rare even in physics: Nancy Cartwright, *How the Laws of Physics Lie* (Oxford: Oxford University Press, 1983). On biology, see John Beatty, "The Evolutionary Contingency Theory," in *Concepts, Theories and Rationality in the Biological Sciences*, ed. G. Wolters and J. G. Lennox (Pittsburgh, PA: University of Pittsburgh Press, 1995), 45–81.

9. Kuhn, *Structure*, 188.

10. Ibid., 190.

11. Renato Dulbecco, "Production of Plaques in Monolayer Tissue Cultures by Single Particles of an Animal Virus," *Proceedings of the National Academy of Sciences U.S.A.* 38 (1952): 747–52.

12. Renato Dulbecco and Marguerite Vogt, "Plaque Formation and Isolation of Pure Lines with Poliomyelitis Viruses," *Journal of Experimental Medicine* 99 (1954): 167–82.

13. Howard M. Temin and Harry Rubin, "Characteristics of an Assay for Rous Sarcoma Virus and Rous Sarcoma Cells in Tissue Culture," *Virology* 6 (1958): 669–88.

14. Howard M. Temin, "The DNA Provirus Hypothesis," *Science* 192 (1976): 1075–80, on 1075. This was Temin's Nobel Prize address; he shared the 1975 prize in physiology and medicine with Renato Dulbecco and David Baltimore.

15. Michel Morange, "From the Regulatory Vision of Cancer to the Oncogene Paradigm, 1975–1985," *Journal of the History of Biology* 30 (1997): 1–29.

16. Frederick L. Schaffer and Carlton E. Schwerdt, "Crystallization of Purified MEF-1 Poliomyelitis Virus Particles," *Proceedings of the National Academy of Sciences U S A* 41 (1955): 1020–23; Wendell M. Stanley, "Isolation of a Crystalline Protein Possessing the Properties of Tobacco-Mosaic Virus," *Science* 81 (1935): 644–45.

17. Angela N. H. Creager and Gregory J. Morgan, "After the Double Helix: Rosalind Franklin's Research on *Tobacco Mosaic Virus*," *Isis* 99 (2008): 239–72, esp. 266–67.

18. Robert E. Kohler, *Lords of the Fly: Drosophila Genetics and the Experimental Life* (Chicago: University of Chicago Press, 1994); Hans-Jörg Rheinberger, *Toward a History of Epistemic Things: Synthesizing Proteins in the Test Tube* (Stanford, CA: Stanford University Press, 1997).

19. For early insights along this line: Karin D. Knorr-Cetina, "The Ethnographic Study of Scientific Work: Towards a Constructivist Interpretation of Science," in *Science Observed: Perspectives on the Social Study of Science*, ed. Karin D. Knorr-Cetina and Michael Mulkay (London: Sage, 1983), 115–40, esp. 118–20.

20. Andrew Pickering, *The Mangle of Practice: Time, Agency, and Science* (Chicago: University of Chicago Press, 1995).

21. Robert E. Kohler, "Systems of Production: *Drosophila, Neurospora*, and Biochemical Genetics," *Historical Studies of the Physical and Biological Sciences* 22 (1991): 87–130; Hans-Jörg Rheinberger, "Experiment, Difference, and Writing. Part 1. Tracing Protein Synthesis. Part 2. The Laboratory Production of Transfer RNA," *Studies in History and Philosophy of Science* 23 (1992): 305–31; 389–422.

22. This more positive view of anomalies is at odds with Kuhn's depiction of science as

basically conservative. Thomas S. Kuhn, "The Essential Tension: Tradition and Innovation in Scientific Research," in *The Essential Tension: Selected Studies in Scientific
Tradition and Change* (Chicago: University of Chicago Press, 1977), 225–39. For an
interesting exchange involving Richard Lewontin, Lorraine Daston, and Carlo Ginz-
burg on whether the notion of "anomalies" even makes sense for biology, see "Editors' Introduction," in *Questions of Evidence: Proof, Practice, and Persuasion across the
Disciplines*, ed. James Chandler, Arnold I. Davidson, and Harry Harootunian (Chicago: University of Chicago Press, 1994), 1–8, on 7.

23. Kohler, *Lords of the Fly*, ch. 2.

24. Drawing on E. P. Thompson as well as work in history of science, Kohler refers to
the norms prevailing in the Drosophilists' community in terms of a "moral economy" of science (ibid., 11–13).

25. Ibid., ch. 7.

26. Muriel Lederman and Richard M. Burian, eds. "The Right Organism for the Job,"
special collection of essays, *Journal of the History of Biology* 26 (1993): 233–67, including Bonnie Claus, "The Wistar Rat as a Right Choice: Establishing Mammalian
Standards and the Idea of a Standardized Mammal," *Journal of the History of Biology* 26 (1993): 329–49; Rachel Ankeny, "The Conqueror Worm: An Historical and
Philosophical Examination of the Use of the Nematode *C. elegans* as a Model Organism" (Ph.D. diss., University of Pittsburgh, 1997); Karen A. Rader, *Making Mice:
Standardizing Animals for American Biomedical Research, 1900–1955* (Princeton, NJ:
Princeton University Press, 2004); Carrie Friese and Adele E. Clarke, "Transposing
Bodies of Knowledge and Technique: Animal Models at Work in Reproductive Sci-
ences," *Social Studies of Science* 42 (2011): 31–52.

27. Ilana Löwy and Jean-Paul Gaudillière, "Disciplining Cancer: Mice and the Practice
of Genetic Purity," in *The Invisible Industrialist: Manufactures and the Production of
Scientific Knowledge*, ed. Jean-Paul Gaudillière and Ilana Löwy (London: Macmillan,
1998), 209–49.

28. Karen A. Rader, "Of Mice, Medicine, and Genetics: C. C. Little's Creation of the Inbred Laboratory Mouse, 1909–1918," *Studies in History and Philosophy of the Biological and Biomedical Sciences* 30C (1999): 319–43, esp. 321–23, and Rader, *Making
Mice*, ch. 1. There has been excellent work along these lines recently: Friese and
Clarke, "Transposing Bodies" and Nicole C. Nelson, "Modeling Mouse, Human,
and Discipline: Epistemic Scaffolds in Animal Behavior Genetics," *Social Studies of
Science* 43 (2012): 3–29.

29. Rheinberger, *Toward a History of Epistemic Things*; and, for a broader account of this
approach, Rheinberger, *An Epistemology of the Concrete: Twentieth-Century Histories of
Life* (Durham, NC: Duke University Press, 2010).

30. Gaudillière and Löwy, *The Invisible Industrialist*. On this point, see also Friese and
Clarke, "Transposing Bodies."

31. Rheinberger, *Toward a History of Epistemic Things*, 28; François Jacob, *The Statue
Within: An Autobiography*, trans. Franklin Phillip (New York: Basic Books, 1988), 9.

32. Harrison Echols, *Operators and Promoters: The Story of Molecular Biology and Its Creators*, ed. Carol A. Gross (Berkeley: University of California Press, 2001), 60–62.

33. Several scholars have been interested in how scientific (and especially biomedical)
research connects different social worlds: Joan H. Fujimura, "Constructing Doable
Problems in Cancer Research: Articulating Alignment," *Social Studies of Science* 17
(1987): 257–93; Susan Leigh Star and James R. Griesemer, "Institutional Ecology,
'Translations' and Boundary Objects: Amateurs and Professionals in Berkeley's

Museum of Vertebrate Zoology, 1907–1939," *Social Studies of Science* 19 (1988): 387–420; Adele E. Clarke and Elihu Gerson, "Symbolic Interactionism in Science Studies," in *Symbolic Interactionism and Cultural Studies*, ed. Howard S. Becker and Michael McCall (Chicago: University of Chicago Press, 1990), 170–214; Ilana Löwy, "The Strength of Loose Concepts—Boundary Concepts, Federative Experimental Strategies and Disciplinary Growth: The Case of Immunology," *History of Science* 30 (1992): 371–96; Jean-Paul Gaudillière, "Oncogenes as Metaphors for Human Cancer: Articulating Laboratory Practices and Medical Demands," in *Medicine and Social Change: Historical and Sociological Studies of Medical Innovation*, ed. Ilana Löwy (John Libbey Eurotext, 1993), 213–47.

34. Rheinberger, *Toward a History of Epistemic Things*, 135.

35. For one detailed example of this point, see Angela N. H. Creager and Jean-Paul Gaudillière, "Meanings in Search of Experiments and Vice-Versa: The Invention of Allosteric Regulation in Paris and Berkeley, 1959–1968," *Historical Studies in the Physical and Biological Sciences* 27 (1996): 1–89.

36. Rheinberger's differentiation of "epistemic object" from experimental system provides one way to tame this chronic conflation between object and system, although "epistemic object" is similarly both material and representational (Rheinberger, *Toward a History of Epistemic Things*).

37. Annette W. Coleman, Lynda J. Goff, and Janet R. Stein-Taylor, eds., *Algae as Experimental Systems* (New York: Alan R. Liss, 1989); James Hanken, "Model Systems versus Outgroups: Alternative Approaches to the Study of Head Development and Evolution," *American Zoology* 33 (1993): 448–56; Elizabeth A. Kellogg and H. Bradley Shaffer, "Model Organisms in Evolutionary Studies," *Systematic Biology* 42 (1993): 409–14; Daniel E. Koshland, "Biological Systems," *Science* 240 (1988): 1385; and Ray White and C. Thomas Caskey, "The Human as an Experimental System in Molecular Genetics," *Science* 240 (1988): 1483–88.

38. Kohler, "Systems of Production," 127.

39. One might include even low-level description as part of a model system. Rachel Ankeny argues that the wiring diagram, specifying all the connections between neurons in the nematode worm, is itself part of the model system. This wiring diagram is not based on one particular worm, but is rather a canonical nervous system based on the neurological connections observed in many standard specimens (R. A. Ankeny, "Model Organisms as Models: Understanding the 'Lingua Franca' of the Human Genome Project," *Philosophy of Science* 68 [2001]: S251–S261).

40. Joshua Lederberg, "*Escherichia coli*," in *Instruments of Science. An Historical Encyclopedia*, ed. Robert Bud and Deborah Jean Warner (New York: Garland Publishing, 1998), 230–32, on 231.

41. A countervailing practice of comparative analysis in twentieth-century life science is neglected in the attention to dominant model systems: Bruno J. Strasser and Soraya de Chadarevian, "The Comparative and the Exemplary: Revisiting the Early History of Molecular Biology," *History of Science* 49 (2011): 317–36. For scientific criticisms of the assumption of typicality in the field of developmental biology, see Jessica Bowker, "Model Systems in Developmental Biology," *Bioessays* 17 (1995): 451–55.

42. Angela N. H. Creager, "Mobilizing Biomedicine: Virus Research Between Lay Health Organizations and the U.S. Federal Government, 1935–1955," in *Biomedicine in the Twentieth Century: Practices, Policies, and Politics*, ed. Caroline Hannaway (Washington, DC: IOS Press, 2008), 171–201; Buhm Soon Park, "Disease Categories and Scientific Disciplines: Reorganizing the NIH Intramural Program, 1945–1960," in

Hannaway, *Biomedicine in the Twentieth Century*, 27–58. Also see "Model Organisms for Biomedical Research," National Institutes of Health, http://www.nih.gov/science/models/.

43. Jean-Paul Gaudillière illustrates the historical complexities of developing animal models for human cancer, particularly for its viral transmission, in "NCI and the Spreading Genes: About the Production of Viruses, Mice and Cancer," in *The Practices of Human Genetics: Sociology of the Sciences Yearbook*, ed. Michael Fortun and Everett Mendelsohn (Dordrecht, The Netherlands: Kluwer Academic Publishers, 1999), 89–124.

44. Angela N. H. Creager and Jean-Paul Gaudillière, "Experimental Arrangements and Technologies of Visualization: Cancer as a Viral Epidemic (1930–1960)," in *Heredity and Infection: The History of Disease Transmission*, ed. Jean-Paul Gaudillière and Ilana Löwy (London: Routledge, 2001), 203–41.

45. Jean-Paul Gaudillière, "The Molecularization of Cancer Etiology in the Postwar United States: Instruments, Politics and Management," in *Molecularizing Biology and Medicine: New Practices and Alliances, 1910s-1970s*, ed. Soraya de Chadarevian and Harmke Kamminga (Amsterdam: Overseas Publishing Association for Harwood Academic Publishers, 1998), 139–70.

46. Wendell M. Stanley, "Viruses and Cancer," lecture presented at the Scientific Session of the American Cancer Society, New York City, October 22, 1962, Stanley papers, carton 20, folder "Manuscripts for talks," Bancroft Library, University of California, Berkeley.

47. Doogab Yi, "The Enemy Within? Oncogenes and the Demise of the Special Virus Cancer Program in the 1970s," paper presented at "Debating Causation: Risk, Biology, Self, and Environment in Cancer Epistemology, 1950-2000," Princeton University, October 21, 2011; Robin Scheffler, "Managing the Future: The Special Virus Leukemia Program and the Acceleration of Biomedical Research," *Studies in History and Philosophy of Biological and Biomedical Sciences* 48 (2014): 231–49.

48. Robin Scheffler, "From Polio to p53: The Life of Simian Virus 40," paper presented at the History of Science Society Meeting, San Diego, November 16, 2012.

49. See Susan *Wright, Molecular Politics: Developing American and British Regulatory Policy for Genetic Engineering, 1972–1982* (Chicago: University of Chicago Press, 1994).

50. I owe this lucid formulation of three criteria for good biological exemplars to Lorraine Daston.

51. Though this topic has been amply treated by others, especially David Kaiser and Andrew Warwick, "Kuhn, Foucault, and the Power of Pedagogy," in *Pedagogy and the Practice of Science: Historical and Contemporary Perspectives*, ed. David Kaiser (Cambridge, MA: MIT Press, 2005), 393–409.

NINE

Structure as Cited, *Structure* as Read

ANDREW ABBOTT

To say something new about Kuhn's book is beyond my power. But like everyone else who has said that sentence, I shall nonetheless go on to say my little bit about this book that has been cited more than once a day since its publication. I speak as an outsider. Although I took field examinations in the sociology of science, the field soon made moves I found uncongenial, so I turned toward the sociology of professions. I never published my own ethnography of a knowledge system—perhaps the psychiatrists were less noble game than the physicists and the biologists. So I have mainly viewed the sociology and history of science from a distance, as one more case for my emerging analysis of disciplines and intellectual camps.

I undertake here a brief, three-part examination of the phenomenon of Kuhn. I begin with a short sketch of his overall impact, using Web of Science (WOS) citation data. I then analyze the page citation data, uncovering some changes in the practices of scholarly reading. I turn finally to my own reading, for in the end, the only proper preparation for such an essay is to read a clean, unmarked copy straight through—in my case, for the first time in thirty-nine years. Scholarly reading should not be the quick scan of the screen nor the glancing search for a favorite passage or a forgotten detail, but rather the eager, questioning study one undertook as an undergraduate, when ideas were new and reading had not yet dwindled into an instrumentality or an obligation.

General Patterns

Let me first sketch the book's general influence, using citation data. I start with a caveat: I have no illusions about such data. In a published analysis of a year's worth of citations to my own book on professions, I found that

75 percent of those citations were unnecessary and that more than 50 percent of them appeared in articles whose arguments showed that their authors had either not read or not understood the book they were citing.[1] So citations obviously have major problems. But since they provide at least a general sense of the regions of Kuhn's impact, I shall use them here.

For data I use the WOS databases, because they involve peer-reviewed materials only. Within them, I look only at citations in the Arts/Humanities and Social Sciences Citation Indexes, because those are my areas of substantive interest. And I restrict my attention to citations of the English-language texts of *Structure*, because my second section will concern page citations, and pagination varies in foreign-language editions. (Of course, citations to English editions are in any case the overwhelming majority of citations in the WOS databases.) Finally, I note that all my retrievals have been done in biennial units, since the new WOS date interface seemed at the time to have unresolvable problems at the annual level.

Under these constraints, there were 15,635 citations to *Structure* between its publication in 1962 and November 2012. Citations rose almost linearly from 1963 to 1985, remained flat at about 450 per year until 2000, then fell off a bit, before rising again rapidly after 2005, possibly because of coverage expansion in the underlying WOS databases. To put it simply, *Structure* achieved saturation over about 20 years, and has since maintained a loosely constant position.

The location of these citations, however, has shifted considerably. Here we must recall three things. First, many journals appear in several WOS categories, so area comparisons are somewhat fuzzy. Second—and by contrast—categorizations remain reasonably consistent over time, which means we can judge change within category as long as we remember that the gradual trend of total citations is distinctly upward over this entire period, both because there were more journals and because reference lists grew longer. Third, after 2008, there were drastic expansions in WOS's coverage, so I take that year as my ending point in the following remarks.

Working through the data area by area, one sees quickly that most individual disciplines follow the same pattern as do the overall numbers: an initial period of more or less linearly increasing engagement is followed by a longer period of stability or slow decline. What varies is the slope of the initial run-up and the stability or decline eventually achieved. The one exception is what we might call the applied area, which comprises many fields but from which I have here chosen as representatives the literatures of education, law, management, business, psychiatry, social work, and in-

Total Citations by General Area

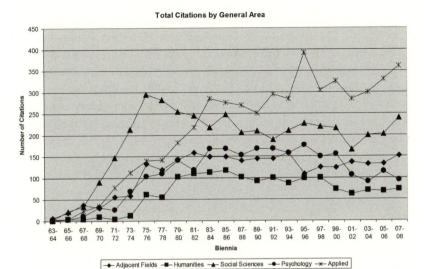

Figure 9.1. Total citations by general area

formation and library science. In these applied areas, taken as a whole, the long-term citation pattern is of steady over-time increase, without any leveling.

As one can see in figure 9.1, knowledge of Kuhn spread first in his immediately adjacent fields: the interdiscipline of history and philosophy of science (HPS) and the disciplines of history and philosophy. This was almost immediately followed in the late 1960s and early 1970s by a broad engagement across the social sciences, led by sociology, a quite unsurprising fact given that the WOS sociology heading includes the sociology of science. Social science citation of Kuhn has however slowly declined from its early peak around 1974, although it remains very substantial.

The humanities—literature, religion, ethics, and so on—discovered Kuhn in a rush after the mid 1970s, achieved saturation by about 1980, and have stayed close to that level since. As for the various psychologies, I have grouped them separately from the social sciences and the humanities. They have followed the same trajectory as the humanities, although there was an eventual fall after 2000.

Finally, the applied fields have taken up Kuhn at a steadily increasing rate. At present, about 40 percent of all citations to Kuhn in this data come in these applied fields— in particular in education and management, which have been the literatures with the most citations to Kuhn since 2000.

Indeed, the education literature has been one of the top three literatures for citation of *Structure* in every biennium since 1969 and was the heaviest citing literature in every biennium from 1985–86 to 1999–2000.

So the overall story is that Kuhn's own fields knew him first and settled quickly into providing a steady 15–20 percent of his citations. The social sciences then took over as the main Kuhn citers after 1970, but settled back to an eventual steady 20–25 percent of citations. Humanities and psychology each had a somewhat later Kuhn expansion—mid- to late 1970s. They eventually settled to about 5 and 10 percent of citations, respectively. Applied fields came to Kuhn gradually through the 1970s and have not yet slaked their thirst for him, providing 40 percent of his current citations. Figure 9.2 shows the results of these trends in terms of the relative percentages of citation of Kuhn at any given time within these broad areas. It is clear that he begins as a figure in his home turf, spends the 1970s and early 1980s as a notable social science citation, and eventually becomes a generic figure widely noticed in applied fields.

Within these particular areas, however, there are some striking differences. Thus, within Kuhn's home turf, philosophy itself engaged Kuhn

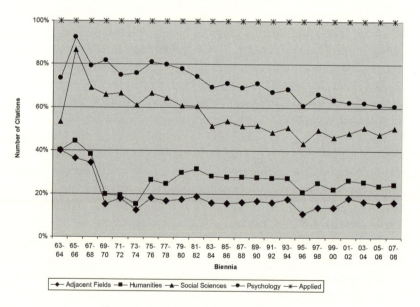

Figure 9.2. Percentage of citations by general area

slowly but made up for lost time from 1974 to 1980, reaching 40 citations a year by 1982. But there has been a steady decline since. Kuhn's own field, HPS, by contrast has shown a long and steadily rising interest that is only now leveling off at about 30 citations per year. In sharp contrast, Kuhn has never had much of a presence in history more generally. Nonetheless, taken altogether, Kuhn's home disciplines steadily provide 75 of his 450 citations each year.

In the social sciences, as I noted, the dominant force is sociology, which went over 50 citations a year by 1974, then fell rapidly back to around 20 or so after 1982. Economics has also been a consistent citer of Kuhn, reaching a plateau of 20 citations per year by 1976 and staying there since. Political science had a distinct burst of interest in the mid 1970s, but has fallen back to 10 to 15 citations a year. Anthropology also had an early burst of interest, but has fallen even further back; Kuhn now appears only 5 to 8 times a year in anthropology journals. Across the social sciences, this adds up to a rapid crescendo of social science interest (to over 150 citations per year) through the mid 1970s and a slower fall since that time.

In the humanities, as I noted, the encounter with Kuhn was later but quick and has nearly maintained its original peak. Waning interest in literature was balanced with waxing interest in religious studies and ethics. Even so, humanities citation of Kuhn is only about one-third to one-half that in the social sciences at any given time. One surmises that not Kuhn but Foucault is serving the humanists as the generic citation for the concept that C. I. Lewis once called "temporary *a prioris*."[2] This theory that Kuhn is essentially a generic citation for the fact that there are successive differences in worldviews is also confirmed by the disinterest of anthropologists, who of course have their own vast literature on that topic, a literature long predating Kuhn.

Among the applied fields, there are some notable trends. Education discovered Kuhn by 1970 and cited him more and more heavily until leveling off around 1992. By contrast, management didn't take up Kuhn until the 1980s and has cited him with increasing frequency ever since. Library and information science discovered him early and has slowly but steadily increased its citation of Kuhn. By contrast with all three of these fields, law and psychiatry both follow a boom-and-bust pattern: a Kuhn vogue in the 1980s and 1990s, much bigger in law, as it happens, but then a sharp fall since, paralleling the trends in the social sciences, but later.

How then should we summarize this data? It seems to me to imply that Kuhn has three major audiences. The first lies in his home fields of philosophy, history and philosophy of science, and sociology of science. In

HPS, the engagement has been early and longstanding. By contrast, both philosophy and the sociology of science engaged him quite heavily and quite early, but eventually began to move on to other things.

The second major audience lies in the social sciences, perhaps including psychology as well. Most of these fields had some major engagement with Kuhn between 1968 and 1980, and all of them other than anthropology retain considerable interest in the book. One guesses that the Kuhnian model of science led to a period of serious self-questioning in most of these fields—it certainly did in sociology[3]—but that after that confrontation, the book has lived on mainly as a generic citation for the fact that views of knowledge do in fact change steadily over time.

The third audience lies in the applied fields. I do not know much about the deployment of Kuhn in these areas. One wonders whether the management literature is not just citing Kuhn to consecrate the latest fad by calling it a new paradigm. One wonders whether information scientists are not using Kuhn as a bludgeon to defeat those who would deny their new and rather vacuous theory of knowledge. One wonders particularly why psychiatrists should include so many longstanding fans of Kuhn: was this perhaps related to the attack on psychoanalysis?

But these must remain speculations. I can say only that there is evidence: first, that Kuhn's actual model of intellectual change still commands attention in his heartland fields; second, that the book has settled back into more generic citation status in most of the social sciences; and third, that the applied literatures are using him very heavily, although in ways we can't identify without further detailed research.

Kuhn Page by Page

We can focus this analysis by considering which parts of Kuhn actually get cited. Let me turn then to a detailed analysis of the page citations to *Structure*.

The book has 210 pages in the long-familiar Chicago edition with the sober typeface and sans-serif running heads of the *International Encyclopedia of Unified Science*.[4] In the most recent, fiftieth-anniversary edition, the pagination has changed slightly, deviating by a page at the postscript, and by two pages at the end.[5] Luckily, that edition has had no time to contribute to WOS citation data. So we have consistent pagination for fifty years. For the record, I retrieved these page citations on November 6, 2012.

As I noted earlier, there have been 15,635 papers citing Kuhn in the humanities and social sciences since 1962. There have been 18,262 days in that time, so the citation rate is impressive: close to once a day in these fields alone. But of these 15,635 citations, a mere 869—around 6 percent—refer to a particular page or chapter. This number is actually an upper bound, since the WOS format makes it extremely laborious to remove duplication from this data: a given paper can appear twice in this number if it cites two different pages of *Structure*. So the actual number of different papers making page citations to Kuhn is probably below 800.

This lack of page citations immediately reveals something quite important. Detailed disciplinary surveys done by the University of Chicago Library School in the early 1950s tell us that as of that time, in the social sciences, about two-thirds of all citations contained page references.[6] In those days, citation was a specific reference to a specific statement, not a general reminder or a generic gesture. Today, of course, the situation is quite different. Most citations in social science journal articles designate no pages. For example, of the two thousand citations to my own first book, published twenty-four years ago, only 3.7 percent contain any page reference. Or again, a glance at the articles in the most recent issues of the two leading sociological journals reveals that only about 5 percent of all citations have page information.

We might then suspect that the rate of page citation of *Structure* has similarly declined. This prediction turns out to be correct. Looking at each biennium since 1963, the percentage of citations that include page information started at around 14 percent, but declined to about 8 percent by the mid 1970s, then to 5 percent by the late 1980s, and finally to 3 percent by the mid 2000s. Interestingly, however, there are occasional bursts of page citation interrupting this decline from time to time, probably caused by a small number of papers—perhaps from a conference—that have gone back to Kuhn in detail. Figure 9.3 shows these data, moving averaged across three biennia (hence across six full years) in order to emphasize the trend.

This longstanding pattern of decline suggests that the vast majority of references to the book is general and vague, not pointed and specific. The page citation data, that is, tend to confirm my earlier hypothesis that the book has become a generic citation for the idea that worldviews change rather than a specific citation signaling a particular theory of how they change.

We might further focus this analysis by asking what parts of *Structure* are most cited among these paltry 869 page citations. It turns out that the first pages of sections dominate: precisely half of all page citations are to the

6yr MA % page citations

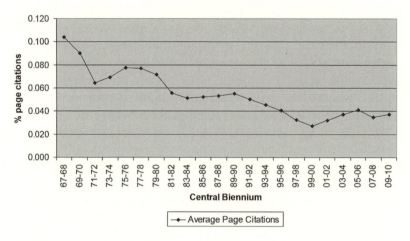

Figure 9.3. Six-year MA % citations

first pages of sections, and, conversely, with minor exceptions, these first pages are cited more than any other pages in the book. The exceptions are somewhat mystifying: the first page of section 9 (on the nature and necessity of revolutions) has been cited only once, and that of section 4 (on normal science as puzzle solving) has been cited only twice. Moreover, the second page of the postscript, with its disentanglement of the various senses of the word "paradigm" has been cited slightly more (22 times) than the incipits of sections 10 (21), 11 (15), and 12 (12) (and a few other pages are tied with section 12).

All the same, it should come as no surprise that the most cited single page is the opening page of section 2, with its definitions of the concepts of normal science and paradigm. This page claims 12 percent of all page citations, although I should remind you that given the rarity of page citation generally, this most important page still represents only 0.6 percent of the total citations of the work.

Taken as whole, page citations are distributed surprisingly well across the chapters. True, section 2 and the postscript each have about 16 percent of the total, but sections 1 (on history), 3 (on normal science), and 12 (on the resolution of revolutions) command about 9 percent apiece. And with the exception of the hapless section 4 (on puzzle-solving), all the rest have from 4 percent to 7 percent.

It is easy to summarize these findings, which are obvious enough. All but a tiny handful of citations are to the whole book. As for that handful, half is to the incipits of the next largest set of units, the sections. And taken as wholes, the two leading sections command about one-third of the total. At each level, citation is very narrowly stacked—a fractal pattern.

As for the distribution of pages by numbers of citations, it's the usual gamma-distributed curve with a huge left-hand bubble and a long, flat tail. I have already talked about the tail: the section incipits with their lion's share of the citations. What about the bubble at the left-hand end? A full 28 percent of the pages have never been cited, and another 23 percent have been cited only once. Many of these uncited pages are devoted to extended examples (e.g., 32–34, 104–107), although occasional example pages—like the discussion of James Clerk Maxwell on 74—have seen several citations. Other uncited pages concern fine points (80–81, 126–128) or ambiguities (139–142).

Let me summarize what I think we learn from the page citation data. First, from the rarity of page citations and their limitation to the core concept pages, I think we can safely guess that the majority of those who have cited this book have not read most—or perhaps even any—of it. The steady decline of page citation over time indicates a declining engagement with the specific arguments of the book and its relegation to the status of generic citation for the notion that there are inevitable changes in world-views, ranging it alongside Foucault, Mannheim, Dilthey, and many others as a standard citation for this commonplace idea.

At the same time, I do not think we can conclude from the fact that 50 percent of the pages are virtually uncited that those pages were not necessary to the book. Without them, *Structure* would not have been more than a large article—it is between fifty thousand and fifty-five thousand words by my estimate, about twice the length of a very long social science article. Articles, I would argue, lack the heft of books. Those fifty largely uncited pages are what dignifies *Structure* as a monograph, capable of the practices that monographs can sustain: multiclass reading assignments, detailed critique of argument, and so on.

All the same, the concentration of page citation on the most easily cited and most obvious places (most notably on the book as an undifferentiated whole) does indicate a possible secular decline in the quality of scholarly reading, paralleling the decline shown by comparison of today's page citation figures with those in the earlier-quoted, discipline-specific citation studies from the mid-1950s.

Reading Kuhn

This sad conclusion about reading brings me to a third topic, my own reading of the work. At 9:35AM on Saturday, November 10, 2012, in a room filled with ten graduate students attending my monthly write-in, I sat down and read *Structure* straight through. It took most of the day. I read a fair copy and made no marks. I did, however, take notes.

As you can guess, this attempt at a pristine, undergraduate reading failed miserably. The many tangents and irrelevancies in my notes betray the quixotic nature of such a project, witnessing the many possible frames that now jostled to control my current reading of the book. I could read it as primary data on the intellectual milieu of its writing, and so indeed I noted the frequent use of gestalt arguments, as dated today as Coke in a bottle. I could read it for its logic and coherence, and so indeed I noted the ad hoc nature of the argument about rules when compared with the smooth flow of the opening argument about normal science and paradigms. I could read it as a case study in the qualities that lead to high citation, and so indeed I noted its brevity and clarity, its focus on a single new concept and a few clear examples.

But in the end, I inevitably reacted most strongly to that aspect of Kuhn's argument that is closest to my own current concerns: his discussion of ideals for knowledge. So I will speak briefly to that topic.

Kuhn's normative views are plain in his closing section on scientific progress. He first asks why science progresses when art, political theory, and philosophy do not. He notes the danger of tautology here: that we might simply use the name "science" for any form of knowledge that has the quality of progressing. He also notes that such progress does appear in nonscientific fields, but only within their subunits or schools, not across those fields as wholes. By contrast, sciences tend not to have equal, competing schools at any given time, but rather successions of dominant schools—that is, paradigms. He reemphasizes that paradigms reduce the cognitive load imposed by always reasoning from first principles and notes how certain institutions contribute to efficiency in scientific production. In particular, he mentions insulation from the larger society and education via textbooks rather than contemplation of actual research. These maximize the production of new knowledge through normal science.

As for progress through paradigm change, Kuhn concedes immediately that the winners write history (so all change becomes "progress" definitionally) but argues that scientific knowledge change is distinguished from sheer "might makes right" by the particular nature of these scientific communities, which alone make the authoritative decision to change paradigm. He notes that scientific communities are communities of problem solvers, that scientists work at a detailed level, that their solutions must be accepted by a community of peers, and that the peers are regarded as sole arbiters. Taken together, these community rules mean that both the list of problems solved and the precision of the individual solutions will grow steadily.[7] What these qualities do not guarantee, however, is that this process necessarily brings us nearer to the truth.

Now Kuhn's successors carefully demolished this specialness claimed for scientific communities, and it is gratifyingly amusing that they did so in a manner that so well exemplified precisely the argument they were at such pains to reject. But it is less interesting to wonder whether Kuhn was mistaken in believing science to possess this special form of social organization than it is to wonder why he did not see that other bodies of culture had quite the same special form. Classical music over the harmonic period had such an organization, and, much more important, the common law had already had such an organization for three centuries by the time Bacon wrote the *New Atlantis*. Moreover, the practitioners of both harmonic music and the common law would have believed in progress in exactly the same sense in which Kuhn deploys that term. To his credit, he does make this point by discussing Renaissance painting early in the chapter, but he then immediately withdraws it by arguing that painting was regarded as cumulative only during the period in which it was close to science.

But regardless of whether there are other forms of knowledge that possess the special social organization that produces what Kuhn calls progress, there remains the inverse problem that there are many important forms of knowledge—and most of the humanities and social sciences are among them—that, while they are disciplined, rigorous, and peer-governed fields, cannot be said to progress. My own current puzzle is how to establish ideals for such knowledge systems.

Suppose we grant Kuhn his linear science with progressing paradigms. Suppose we grant him that science creates more—and more precise—problem-solving over time. What should be ideal temporal trajectories for fields that have contending schools within which there is "progress" of the Kuhnian sort, but which cannot or do not envision an overall progress, represented in a hegemonic paradigm? Are such fields doomed to mere succession of fashions? This is a very real question, for most of us here are practitioners of such fields: historians, philosophers, sociologists, and so on. We know perfectly well that cumulation does not really exist in our fields; that much of today's hot research is rediscovering or restating or relabeling ideas familiar fifty or a hundred or a hundred and fifty years ago—old arguments dressed up in new data, new methods, new citations.[8]

Yet we also believe in disciplined inquiry, in careful training of graduate students, in the elaboration of painfully exact and often quite "scientific" methodologies. We believe in rigor.

But what do we expect rigor to produce? We don't know, for we have no normative conception for how our disciplines ought to evolve over time, if progress is ruled out as a normative goal. And this is not simply an internal question. It is also relevant to fights with external "knowledge providers" who would be happy to replace us. I have found this out most clearly when arguing against Google and its fellows by questioning whether speed and ease of library work would necessarily improve library research. In the middle of this polemic, I suddenly realized that my real problem was not understanding how speed and ease affected research work, important as that is.[9] It was rather that I had no criterion by which to judge whether that work had in fact been improved or damaged, by Google or anything else. I knew intuitively that there was no sense in thinking that if we could do our library work twice as fast there would result huge advances in English literature, say, or historical sociology or philosophy. Speed mattered only if one believed that there were "true" answers to be found—an ultimate perfect interpretation of *Pride and Prejudice*, say, or a final definition of the notion of virtue. And of course these do not exist. But that doesn't mean that we shouldn't do research and writing on these topics, and moreover that we shouldn't do them in disciplined ways evaluated by peers. Yet what then should be the ideals for such a system? What should it accomplish?

I shall specify such ideals in terms of a space of possible knowledge. That possibility space is generated by the combinations of the various attributes of knowledge: things like categorical forms of thinking (individualism versus emergentism in the social sciences, for example), as well as preference for certain genres of writing and for types of examples or data. Such attributes are all the constituents of Kuhn's paradigms, in short, but seen as varying properties of local schools rather than of hegemonic paradigms. Understood as continuous dimensions, these attributes can be seen as a basis that generates the space of all possible forms of knowledge. (If the attributes be understood as discrete qualities, the space of all possible forms of knowledge can be defined as the power set of all combinations of these attributes of knowing.)

The idea that there is no progress is the idea that there is no normatively preferable trajectory through this space. I take that for granted, at least with respect to the humanities and most of the social sciences. What then are alternative ideals? Surely, simple random change is not a viable candidate!

One possible nonprogressive knowledge ideal is that no region of that space should disappear from our repertoire of current knowing for very long. In formal terms, we ask for an upper bound on the mean recurrence time for any substantial region of the knowledge space. For example, we don't want individualist accounts of social life to so dominate that there are no structural or corporate ones for more than, say, a decade. We don't want symbolic interpretations to be so dominant that social structure disappears from the social science agenda for too long. We don't want externalist literary criticism to drive internalism permanently from the scene.

Another normative criterion for a noncumulative knowledge process is plenitude. By plenitude, I mean that we want a knowledge process that will effectively fill the space of possible knowledges of the social world. A trivial way of imagining that is to think that we would like ultimately to investigate all possible ways of knowing, in the sense of visiting all possible combinations of all possible attributes of knowing; ethnographic narrative, positivism, quantitative lyrical conflict-based prosopography, and so on. But this is a static conception, and it presupposes that we already know all possible dimensions of attributes and that all are invoked at any given time. Plenitude should include the possibility of making new dimensions of difference or of eradicating old ones, and we should envision rules for producing that kind of invention.

But at any given time, we should aim for maximal current plenitude in nonprogressive knowledge. So, for example, it is a good thing that today's economists are reinventing experimental psychology, not because a) "the psychologists didn't do it right and thank goodness the economists are now there to show them how to do real mathematics" (which would be the cumulational way to think about this), nor because b) doing social psychological experiments will revolutionize economics (it won't), but because somebody needs to be doing this kind of thing and the research front of psychology has in fact gone biological for the moment, leaving this whole area of experimental approaches to behavior unexpressed in the world of social science. Meanwhile, the sociologists and political scientists have taken up large-scale punditry about the economy, since the economists have become so mathematical that actual philosophical reflection about economic life has largely disappeared from their agenda. In both cases, what appear to be reduplications and reinventions are actually part of maintaining at any given time the plenitude of the social sciences as ways of thinking about the social world.

A final general criterion arises from the fact that a given set of ideas can be discovered in several different ways, from several different backgrounds. To continue my example, although the economists are reinventing experimental psychology, they are doing it differently, perhaps more theoretically, or perhaps with a different theoretical approach, or perhaps with more formal—if not necessarily more correct—statistics. Political scientists and some sociologists are of course also in the experimental business these days, and they are doing the same kinds of things with still different theoretical assumptions, different experimental protocols, different ways of choosing subjects, and so on. It is clearly better for social science as a whole to have these many different approaches to the same set of issues. Not because the variety will lead us to the correct answer faster—there is no correct answer. But rather because it is simply better to have thought the same thing in many different ways.

I shall call this the plurality criterion. We may state it as follows: it is better for nonprogressive forms of knowledge to have more routes to the same general area in the space of knowledge than to have fewer. The plurality criterion is a way of embracing the fact that most new production in nonprogressive knowledge is in some sense rediscovery. Watching physicists rediscover network analysis has provided many of us sociologists with considerable amusement recently, but it is still a good thing to have a new voice in that area. It may sometimes be an arrogant and contemptuous voice, but it's nonetheless enriching to our studies. The same is true for my brilliant colleague Steve Levitt's *Freakonomics*, most of which seemed to be warmed-over sociology tarted up with clever mathematics to appear to be revolutionary economics. On a larger scale, the more solid parts of cultural studies are reinventing classical anthropology, to the intense rage of those anthropologists old enough to have read the classics. But it's nonetheless important to have those ideas refurbished for a new generation. The younger anthropologists, sadly, seem to be busy forgetting them.

In summary, I propose three basic knowledge ideals for nonprogressive forms of knowledge: bounded recurrence, plenitude, and plurality. Note that because nobody can know everything, we are unable to stand outside the system of knowledge and to know, at any given time, whether these ideals are being met. However, we can perhaps infer how scholarly communities ought to behave locally if these criteria are to be maximized. We can't know to what extent we have succeeded, to be sure. But we can at least so act locally that success is maximally likely.

But this requires specification of local rules, for which I do not have time here. I can merely say that, once again, reading Kuhn confronts us with crucial issues. By discussing the temporal ideal that is progress, Kuhn implicitly challenges us to provide the alternative temporal ideals that should govern nonprogressive knowledge. It is up to us to rise to that challenge.

Notes

1. Andrew Abbott, "Varieties of Ignorance," *American Sociologist* 41 (2010): 174–89.

2. Clarence Irving Lewis, *Mind and the World Order* (Chicago: Scribners, 1929), c. 8. I confess that I cannot find this exact term in Lewis's text, although the concept of a temporally changing a priori is the subject of an entire chapter of his book. When I was preparing this talk, I found the remark "just like Lewis's temporary a priori" in my own handwriting on the last page of my old copy of *Structure* from the 1970s, and I assumed without checking that the phrase was Lewis's and should therefore be attributed. But on careful inspection, I could not find it when I was finishing this publication version. Perhaps it was my own summary of Lewis. But it was certainly Lewis's idea, not mine, so I have let the text stand as originally I delivered it. Writing in the late 1920s, Lewis argued quite persuasively for a general drift of the a priori over time, a drift of which something like paradigmatic change would be a subset. Indeed, the chapter sets forth concepts that are quite close to paradigms and paradigmatic incommensurability, although Lewis's argument does not include the "thought collective" notion of Fleck and Kuhn. I have little idea how I even came to own Lewis's book. I must have picked it up randomly in Powell's Bookstore one day and, finding it filled with a priori, "probability," and all those other words that then interested me, must have thought it cheap at $1.50. My copy is very carefully annotated, and it is clear, on rereading, that this book influenced me greatly, for I read it—as the note above shows—before I read Kuhn.

3. E.g., George Ritzer, *Sociology: a Multiple Paradigm Science* (Boston: Allyn and Bacon, 1975).

4. Thomas Kuhn, *The Structure of Scientific Revolutions*, 2nd ed. (Chicago: University of Chicago Press, 1970). When page numbers are given in the text or *Structure* is cited in a note, this is the edition to which I refer.

5. Thomas Kuhn, *The Structure of Scientific Revolutions*, introduction by Ian Hacking. 4th ed. (Chicago: University of Chicago Press, 2012).

6. Andrew Abbott, "Library Research Infrastructure for Humanistic and Social Scientific Scholarship in America in the Twentieth Century," in *Social Knowledge in the Making*, ed. C. Camic, N. Gross, and M. Lamont, 43–87 (Chicago: University of Chicago Press, 2011), 70.

7. Kuhn, *Structure*, 170.

8. I have sketched this argument at more length in the first chapter of *Chaos of Disciplines* (Chicago: University of Chicago Press, 2001). A more recent discussion, from which much of this argument is drawn, is "The Vicissitudes of Methods," Plenary Lecture, Economic and Social Research Council 5th Annual Methods Festival, Oxford, July 3, 2012.

9. Andrew Abbott, "Googles of the Past: Concordances and Scholarship," *Social Science History* 37 (2013): 427–55.

BIBLIOGRAPHY

Abbott, Andrew. "Googles of the Past: Concordances and Scholarship." *Social Science History* 37 (2013): 427–55.
———. "Library Research Infrastructure for Humanistic and Social Scientific Scholarship in America in the Twentieth Century." In *Social Knowledge in the Making,* edited by C. Camic, N. Gross, and M. Lamont, 43–87. Chicago: University of Chicago Press, 2011.
———. "Varieties of Ignorance." *American Sociologist* 41 (2010): 174–89.
———. "The Vicissitudes of Methods." Plenary Lecture, Economic and Social Research Council 5th Annual Methods Festival, Oxford, July 3, 2012.
Andresen, Jensine. "Crisis and Kuhn." *Isis* 90 (1999): S43–S67.
Ankeny, Rachel. "The Conqueror Worm: An Historical and Philosophical Examination of the Use of the Nematode *C. elegans* as a Model Organism." Ph.D. diss., University of Pittsburgh, 1997.
———. "Model Organisms as Models: Understanding the 'Lingua Franca' of the Human Genome Project." *Philosophy of Science* 68 (2001): S251–S261.
Aristotle. *The Complete Works of Aristotle. The Revised Oxford Translation.* Edited by J. Barnes. 2 vols. Princeton, NJ: Princeton University Press, 1984.
Arthos, John. "Where There Are No Rules or Systems to Guide Us: Argument from Example in a Hermeneutic Rhetoric." *Quarterly Journal of Speech* 89 (2003): 320–34.
Ash, Mitchell. *Gestalt Psychology in German Culture, 1890–1967: Holism and the Quest for Objectivity.* New York: Cambridge University Press, 1995.
Bachou, Jean. "Discours a la recommandation de la Philosophie ancienne restablie en sa pureté ; Et sur le nom de son premier Autheur." Preface to *La Philosophie naturelle restablie en sa pureté,* by Jean d'Espagnet. Translated by Bachou. Paris: Edme Pepingué, 1651.
Bacon, Francis. *The Instauratio magna Part II:* Novum organum *and Associated Texts.* Edited by Graham Rees with Maria Wakeley. Vol. 11 of *The Oxford Francis Bacon.* Oxford: Oxford University Press, 2004.
———. *Philosophical Studies c. 1611—c. 1619.* Edited by Graham Rees. Vol. 6 of *The Oxford Francis Bacon.* Oxford: Oxford University Press, 1996.
———. *Sylva Sylvarum or a Natural History in Ten Centuries.* London: W. Lee, 1626.
———. *The Works of Francis Bacon.* Vol. 3. Edited by J. Spedding, R. L. Ellis, and D. D. Heath. London: Longman and Co., 1858–74.

Bakhtin, Mikhail. *The Dialogic Imagination*. Edited by Michael Holquist. Translated by Caryl Emerson and Michael Holquist. Austin: University of Texas Press, 1981.

Barber, Bernard. "Review of *The Structure of Scientific Revolutions*." *American Sociological Review* 28 (1963): 298–99.

Barnes, Barry. *Interests and the Growth of Knowledge*. London: Routledge & Kegan Paul, 1977.

Bartha, Paul. *By Parallel Reasoning: The Construction and Evaluation of Parallel Arguments*. New York: Oxford University Press, 2010.

Barthes, Roland. *Image, Music, Text*. Translated by Stephen Heath. New York: Hill and Wang, 1977.

———. *Sur Racine*. Paris: Editions du Seuil, 1960.

Beatty, John. "The Evolutionary Contingency Theory." In *Concepts, Theories and Rationality in the Biological Sciences*, edited by G. Wolters and J. G. Lennox, 45–81. Pittsburgh, PA: University of Pittsburgh Press, 1995.

Bloor, David. "Left and Right Wittgensteinians." In *Science as Practice and Culture*, edited by Andrew Pickering, 266–83. Chicago: Chicago University Press, 1992.

Bohm, David. "Review of *The Structure of Scientific Revolutions*." *Philosophical Quarterly* 14 (1964): 377–79.

Boring, Edwin G. *A History of Experimental Psychology*. 2nd ed. New York: Appleton, 1950.

———."A New Ambiguous Figure." *American Journal of Psychology* 42 (1930): 444–45.

———. *Sensation and Perception in the History of Experimental Psychology*. New York: Appleton, 1942.

Borradori, Giovanna. *The American Philosopher: Conversations with Quine, Davidson, Putnam, Nozick, Danto, Rorty, Cavell, MacIntyre, and Kuhn*. Chicago: University of Chicago Press, 1994.

Bowker, Jessica. "Model Systems in Developmental Biology." *Bioessays* 17 (1995): 451–55.

Bruner, Jerome. *In Search of Mind: Essays in Autobiography*. New York: Harper and Row, 1983.

Bruner, Jerome, and Leo Postman. "On the Perception of Incongruity: A Paradigm." *Journal of Personality* 18 (1949): 206–23.

Buchwald, Jed Z., and George E. Smith, "Thomas S. Kuhn, 1922–1996." *Philosophy of Science* 64 (1997): 361–76.

Burnyeat, Myles. "Enthymeme: Aristotle on the Logic of Persuasion." In *Aristotle's Rhetoric: Philosophical Essays*, edited by D. Furley and A. Nehamas, 3–55. Princeton, NJ: Princeton University Press, 1994.

———. "Enthymeme: Aristotle on the Rationality of Rhetoric." In *Essays on Aristotle's Rhetoric*, edited by A. Rorty, 88–115. Berkeley: University of California Press, 1996.

———. "The Origins of Non-Deductive Inference." In *Science and Speculation: Studies in Hellenistic Theory and Practice*, edited by J. Barnes et al., 193–238. Cambridge: Cambridge University Press, 1982.

Butterfield, Herbert. *The Origins of Modern Science, 1300–1800*. London: G. Bell, 1949.

Cairns, John, Gunther S. Stent, and James D. Watson, eds. *Phage and the Origins of Molecular Biology*. Cold Spring Harbor, NY: Cold Spring Harbor Laboratory Press, 1966.

Cartwright, Nancy. *How the Laws of Physics Lie*. Oxford: Clarendon Press, 1983.

Chandler, James, Arnold I. Davidson, and Harry Harootunian. Editors' introduction to *Questions of Evidence: Proof, Practice, and Persuasion across the Disciplines*, edited by Chandler, Davidson and Harootunian, 1–8. Chicago: University of Chicago Press, 1994.

Clarke, Adele E., and Elihu Gerson. "Symbolic Interactionism in Science Studies." In *Symbolic Interactionism and Cultural Studies*, edited by Howard S. Becker and Michael McCall, 170–214. Chicago: University of Chicago Press, 1990.

Claus, Bonnie. "The Wistar Rat as a Right Choice: Establishing Mammalian Standards and the Idea of a Standardized Mammal." In Lederman and Burian, "The Right Organism for the Job," 329–49.

Cohen, H. Floris. *The Scientific Revolution: A Historiographical Inquiry*. Chicago: University of Chicago Press, 1994.

Cohen-Cole, Jamie. "The Creative American: Cold War Salons, Social Science, and the Cure for Modern Society." *Isis* 100 (2009): 219–62.

———. "Instituting the Science of Mind: Intellectual Economies and Disciplinary Exchange at Harvard's Center for Cognitive Studies." *British Journal for the History of Science* 40 (2007): 567–97.

———. *The Open Mind: Cold War Politics and the Sciences of Human Nature*. Chicago: University of Chicago Press, 2014.

Coleman, Annette W., Lynda J. Goff, and Janet R. Stein-Taylor, eds. *Algae as Experimental Systems*. New York: Alan R. Liss, 1989.

Collins, Harry M. *Changing Order: Replication and Induction in Scientific Practice*. London: Sage, 1985.

———. *Tacit and Explicit Knowledge*. Chicago: University of Chicago Press, 2010.

Conant, James Bryant. Conant Papers. Harvard University Archives, Cambridge, MA.

———. *Education in a Divided World: The Function of the Public Schools in Our Unique Society*. New York: Greenwood Press, 1948.

———. *Education and Liberty*. Cambridge, MA: Harvard University Press, 1953.

———, ed. *Harvard Case Studies in the History of the Experimental Sciences*. Cambridge, MA: Harvard University Press, 1950.

———. *On Understanding Science*. New Haven, CT: Yale University Press, 1947.

Creager, Angela N. H. *The Life of a Virus: Tobacco Mosaic Virus as an Experimental Model, 1930–1965*. Chicago: University of Chicago Press, 2002.

———. "Mobilizing Biomedicine: Virus Research between Lay Health Organizations and the U.S. Federal Government, 1935–1955." In Hannaway, *Biomedicine in the Twentieth Century*, 171–201.

Creager, Angela N. H., and Jean-Paul Gaudillière. "Experimental Arrangements and Technologies of Visualization: Cancer as a Viral Epidemic (1930–1960)." In *Heredity and Infection: The History of Disease Transmission*, edited by Jean-Paul Gaudillière and Ilana Löwy, 203–41. London: Routledge, 2001.

———. "Meanings in Search of Experiments and Vice-Versa: The Invention of Allosteric Regulation in Paris and Berkeley, 1959–1968." *Historical Studies in the Physical and Biological Sciences* 27 (1996): 1–89.

Creager, Angela N. H., Elizabeth Lunbeck, and M. Norton Wise, eds. *Science without Laws: Model Systems, Cases, Exemplary Narratives*. Durham, NC: Duke University Press, 2007.

Creager, Angela N. H., and Gregory J. Morgan. "After the Double Helix: Rosalind Franklin's Research on Tobacco Mosaic Virus." *Isis* 99 (2008): 239–72.

Daston, Lorraine. "Structure." *Historical Studies in the Natural Sciences* 42 (2012): 496–99.

Dear, Peter. *Discipline and Experience: The Mathematical Way in the Scientific Revolution*. Chicago: University of Chicago Press, 1995.

———. "Science Is Dead; Long Live Science." *Osiris* 27 (2012): 37–55.

De Clave, Étienne. *Paradoxes ou Traittez Philosophiques des Pierres et Pierreries, contre l'opinion vulgaire*. Paris: Pierre Chevaliere, 1635.

Descartes, René. *Oeuvres de Descartes*. Vol. 1. Edited by Charles Adam and Paul Tannery. Paris: J. Vrin, 1996.

Diamond, Cora. *The Realistic Spirit: Wittgenstein, Philosophy, and the Mind*. Cambridge, MA: MIT Press, 1991.

Diderot, Denis, and Jean d'Alembert, eds. *Encyclopédie, ou Dictionnaire raisonné des sciences, des arts et des métiers*. 35 vols. Lausanne/Berne: Les sociétés typographiques, 1751–80.

Douglas, Mary. *Purity and Danger: An Analysis of Concepts of Pollution and Taboo*. New York: Praeger, 1966.

Dulbecco, Renato. "Production of Plaques in Monolayer Tissue Cultures by Single Particles of an Animal Virus." *Proceedings of the National Academy of Sciences U.S.A.* 38 (1952): 747–52.

Dulbecco, Renato, and Marguerite Vogt. "Plaque Formation and Isolation of Pure Lines with Poliomyelitis Viruses." *Journal of Experimental Medicine* 99 (1954): 167–82.

Dupré, John. *The Disorder of Things: Metaphysical Foundations of the Disunity of Science*. Cambridge, MA: Harvard University Press, 1993.

Drake, Stillman. *Discoveries and Opinions of Galileo*. Garden City, NY: Doubleday, 1957.

Echols, Harrison. *Operators and Promoters: The Story of Molecular Biology and Its Creators*. Edited by Carol A. Gross. Berkeley: University of California Press, 2001.

Farrington, Benjamin. *The Philosophy of Francis Bacon: An Essay on Its Development from 1603–1609*. Liverpool, UK: Liverpool University Press, 1964.

Feyerabend, Paul. *Against Method*. London: Verso, 1975.

Fleck, Ludwik. *Enstehung und Entwicklung einer wissenschaftlichen Tatsache. Einführung in die Lehre von Denkstil und Denkkollektiv*. Basel: B. Schwabe, 1935.

Forrester, John. "On Kuhn's Case: Psychoanalysis and the Paradigm." *Critical Inquiry* 33 (2007): 782–819.

Foucault, Michel. *Discipline and Punish*. Translated by Alan Sheridan. New York: Pantheon, 1977.

Frey, Jean-Cecile (Ianus Caecilius). *Cribrum philosophorum qui Aristotelem superiore et hac aetate oppugnarunt*. In *Opuscula varia nusquam edita*, 29–89. Paris: Petrus David, 1646.

Friese, Carrie, and Adele E. Clarke, "Transposing Bodies of Knowledge and Technique: Animal Models at Work in Reproductive Sciences." *Social Studies of Science* 42 (2011): 31–52.

Fujimura, Joan H. "Constructing Doable Problems in Cancer Research: Articulating Alignment." *Social Studies of Science* 17 (1987): 257–93.

Fuller, Steve. *Thomas Kuhn: A Philosophical History for Our Times*. Chicago: University of Chicago Press, 2000.

Gadamer, Hans-Georg. *Truth and Method*. Translated by G. Barden and J. Cumming. London: Sheed and Ward, 1975.

Galen, Claudius. *De temperamentis libri III*. Edited by Georg Helmreich. Leipzig: B. G. Teubner, 1904.

Galison, Peter. "The Americanization of Unity." *Daedalus* 127 (1998): 45–71.

———. *How Experiments End*. Chicago: University of Chicago Press, 1987.

Galison, Peter, and David Stump, eds. *The Disunity of Science: Boundaries, Contexts, and Power*. Stanford, CA: Stanford University Press, 1996.

Garber, Daniel. *Descartes' Metaphysical Physics*. Chicago: University of Chicago Press, 1992.

———. "Galileo, Newton and All That: If It Wasn't a Scientific Revolution, What Was It? (A Manifesto)." *Circumscribere* 7 (2009): 9–18.

———. "Review of Wellman, *Making Science Social.*" *Early Science and Medicine* 10 (2005): 428–34.

Garfield, Eugene. "A Different Sort of Great-Books List: The 50 Twentieth-Century Works Most Cited in the *Arts & Humanities Citation Index, 1976–1983.*" *Current Contents* 16 (1987): 3–7.

Gaudillière, Jean-Paul. "The Molecularization of Cancer Etiology in the Postwar United States: Instruments, Politics and Management." In *Molecularizing Biology and Medicine: New Practices and Alliances, 1910s-1970s*, edited by Soraya de Chadarevian and Harmke Kamminga, 139–70. Amsterdam: Overseas Publishing Association for Harwood Academic Publishers, 1998.

———. "NCI and the Spreading Genes: About the Production of Viruses, Mice and Cancer." In *The Practices of Human Genetics: Sociology of the Sciences Yearbook*, edited by Michael Fortun and Everett Mendelsohn, 89–124. Dordrecht: Kluwer Academic Publishers, 1999.

———. "Oncogenes as Metaphors for Human Cancer: Articulating Laboratory Practices and Medical Demands." In *Medicine and Social Change: Historical and Sociological Studies of Medical Innovation*, edited by Ilana Löwy, 213–47. Montrouge, France, and London: John Libbey Eurotext, 1993.

Gaudillière, Jean-Paul, and Ilana Löwy, eds. *The Invisible Industrialist: Manufactures and the Production of Scientific Knowledge.* London: Macmillan, 1998.

Gillispie, Charles C. "Review of *The Structure of Scientific Revolutions.*" *Science* 138 (1962): 1251–53.

Golinski, Jan. "Is it Time to Forget Science? Reflections on Singular Science and Its History." *Osiris* 27 (2012): 19–36.

Gooding, David, Trevor Pinch, and Simon Schaffer, eds. *The Uses of Experiment: Studies in the Natural Sciences.* Cambridge: Cambridge University Press, 1993.

Goodman, Nelson. "Seven Strictures on Similarity." In *Problems and Projects.* Indianapolis, IN: Bobbs-Merrill, 1972.

Gordin, Michael D. and Erika Lorraine Milam. "A Repository for More than Anecdote: Fifty Years of *The Structure of Scientific Revolutions*," and other essays. *Historical Studies in the Natural Sciences* 42, no. 5 (2012): 476–580.

Gordon, Peter E., et al. "Forum: Kuhn's *Structure* at Fifty." *Modern Intellectual History* 9, no.1 (2012): 73–147.

Hacking, Ian. *The Emergence of Probability.* Cambridge: Cambridge University Press, 1975.

———. Introduction to *The Structure of Scientific Revolutions.* 4th ed. Chicago: University of Chicago Press, 2012.

———. *Representing and Intervening: Introductory Topics in the Philosophy of Science.* Cambridge: Cambridge University Press, 1983.

———. "What Logic Did to Rhetoric." *Journal of Cognition and Culture* 13 (2013): 413–30.

Hall, Maria Boas. "Review of *The Structure of Scientific Revolutions.*" *American Historical Review* 68 (1963): 700–1.

Hanken, James. "Model Systems versus Outgroups: Alternative Approaches to the Study of Head Development and Evolution." *American Zoology* 33 (1993): 448–56.

Hannaway, Caroline, ed. *Biomedicine in the Twentieth Century: Practices, Policies, and Politics.* Washington DC: IOS Press, 2008.

Hanson, Norwood Russell. *Patterns of Discovery.* 2nd ed. Cambridge: Cambridge University Press, 1972.

Hardin, Garrett. "The Competitive Exclusion Principle." *Science* 131 (1960): 1292–97.

———. "The Tragedy of the Commons." *Science* 162 (1968): 1243–48.

Harré, Rom. "Where Models and Analogies Really Count." *International Studies in the Philosophy of Science* 2 (1988): 118–33.

Harris, Steven J. "Introduction: Thinking Locally, Acting Globally." *Configurations* 6 (1998): 131–39.

Heereboord, Adrien. "Consilium de ratione studendi philosophiae." In *Meletemata philosophica*. Vol. 1. Leiden: Franciscus Moyardus, 1654.

Heilbron, J. L. "Thomas Samuel Kuhn, 18 July 1922–17 June 1996." *Isis* 89 (1998): 505–55.

Hershberg, James. *James B. Conant: From Harvard to Hiroshima and the Making of the Nuclear Age*. New York: Knopf, 1993.

Hesse, Mary. *Models and Analogies in Science*. 2nd ed. Notre Dame, IN: Notre Dame University Press, 1966.

———. "Review of *The Structure of Scientific Revolutions*." *Isis* 54 (1963): 286–87.

Holland, Guy. *The Grand Prerogative of Humane Nature Namely, the Souls Naturall or Native Immortality*. London: Roger Daniel, 1653.

Holton, Gerald. "From the Vienna Circle to Harvard Square: The Americanization of a European World Conception." In *Scientific Philosophy: Origins and Developments*, edited by Friedrich Stadler, 47–73. Boston: Kluwer, 1993.

Hook, Sidney. Hook Papers. Hoover Institution Archives, Stanford, CA.

Hoyningen-Huene, Paul. "More Letters by Paul Feyerabend to Thomas S. Kuhn on *Proto-Structure*." *Studies in the History and Philosophy of Science* 37 (2006): 610–32.

———. *Reconstructing Scientific Revolutions: Thomas S. Kuhn's Philosophy of Science*. Chicago: University of Chicago Press, 1993.

———. "Thomas S. Kuhn." *Journal for General Philosophy. Zeitschrift für allgemeine Wissenschaftstheorie* 28 (1997): 235–56.

———. "Two Letters of Paul Feyerabend to Thomas S. Kuhn on a Draft of *The Structure of Scientific Revolutions*." *Studies in History and Philosophy of Science* 26, no. 3 (1995): 353–387.

Hufbauer, Karl. "From Student of Physics to Historian of Science: T. S. Kuhn's Education and Early Career, 1940–1958." *Physics in Perspective* 14 (2012): 421–78.

Isaac, Joel. *Working Knowledge: Making the Human Sciences from Parsons to Kuhn*. Cambridge, MA: Harvard University Press, 2012.

Jacob, François. *The Statue Within: an Autobiography*. Translated by Franklin Phillip. New York: Basic Books, 1988.

Jacobs, Struan. "Michael Polanyi and Thomas Kuhn: Priority and Credit." *Tradition and Discovery: The Polanyi Society Periodical* 33, no. 2 (2006/2007): 25–36.

———. "Thomas Kuhn's Memory." *Intellectual History Review* 19 (2009): 83–101.

Jonsen, Albert R., and Stephen Toulmin. *The Abuse of Casuistry: A History of Moral Reasoning*. Berkeley: University of California Press, 1988.

Kaiser, David. "A Tale of Two Textbooks: Experiments in Genre." *Isis* 103 (2012): 126–38.

Kaiser, David, and Andrew Warwick. "Kuhn, Foucault, and the Power of Pedagogy." In *Pedagogy and the Practice of Science: Historical and Contemporary Perspectives*, edited by David Kaiser, 393–409. Cambridge, MA: MIT Press, 2005.

Kellogg, Elizabeth A., and H. Bradley Shaffer. "Model Organisms in Evolutionary Studies." *Systematic Biology* 42 (1993): 409–14.

Kennedy, G. A. *Aristotle: On Rhetoric. A Theory of Civic Discourse*. 2nd ed. New York: Oxford University Press, 2007.

Knorr-Cetina, Karin D. "The Ethnographic Study of Scientific Work: Towards a Constructivist Interpretation of Science." In *Science Observed: Perspectives on the Social Study of Science*, edited by Karin D. Knorr-Cetina and Michael Mulkay, 115–40. London: Sage, 1983.

Koestler, Arthur. "Arthur Koestler." In *The God that Failed: A Confession*, edited by Richard Crossman. 2nd ed. New York: Columbia University Press, 2001.

Kohler, Ivo. *Über Aufbau und Wandlungen der Wahrnehmungswelt, insbesondere über 'bedingte Empfindungen'*, Österreichische Akademie der Wissenschaften, Philosophischhistorische Klasse. Vol. 227. Vienna: Rudolf M. Rohrer, 1951.

Kohler, Robert E. *Lords of the Fly: Drosophila Genetics and the Experimental Life*. Chicago: University of Chicago Press, 1994.

———. "Systems of Production: *Drosophila, Neurospora*, and Biochemical Genetics." *Historical Studies of the Physical and Biological Sciences* 22 (1991): 87–130.

Koshland, Daniel E. "Biological Systems." *Science* 240 (1988): 1385.

Koyré, Alexander. *Études Galiléennes*. Paris: Hermann, 1939.

Krüger, Lorenz, Lorraine Daston, and Michael Heidelberger, eds. *The Probabilistic Revolution*. Vol. 1. Cambridge, MA: MIT Press, 1987.

Kuhn, Thomas S. "Afterwords." In *World Changes: Thomas Kuhn and the Nature of Science*, edited by Paul Horwich, 311–41. Cambridge, MA: MIT Press, 1993.

———. "An Application of the W. K. B. Method to the Cohesive Energy of Monovalent Metals." *Physical Review* 79 (August 1950): 515–19.

———. "A Convenient General Solution of the Confluent Hypergeometric Equation, Analytic and Numerical Development." *Quarterly of Applied Mathematics* 9 (1951): 1–16.

———. *The Copernican Revolution: Planetary Astronomy in the Development of Western Thought*. Cambridge, MA: Harvard University Press, 1957. Reprint, New York: Modern Library Paperbacks, 1959.

———. "A Discussion with Thomas S. Kuhn." *Neusis: Journal for the History and Philosophy of Science and Technology*, 6 (1997): 143–98.

———. *The Essential Tension: Selected Studies in Scientific Tradition and Change*. Edited by Lorenz Krüger. Chicago: University of Chicago Press, 1977.

———. Foreword to *Genesis and Development of a Scientific Fact*, by Ludwig Fleck. Chicago: University of Chicago Press, 1979.

———. "The Function of Dogma in Scientific Research." In *Scientific Change*, edited by A. C. Crombie, 347–69. New York: Basic Books, 1963.

———. "History of Science." In *International Encyclopedia of the Social Sciences*, edited by D. L. Sills, 13:74–83. London and New York: Macmillan and Free Press, 1972.

———. "Logic of Discovery or Psychology of Research?" In Lakatos and Musgrave, *Criticism and the Growth of Knowledge*, 1–24.

———. "Reflections on My Critics." In Lakatos and Musgrave, *Criticism and the Growth of Knowledge*, 231–78.

———. *The Road since Structure: Philosophical Essays, 1970–1993, with an Autobiographical Interview*. Edited by James Conant and John Haugeland. Chicago: University of Chicago Press, 2000.

———. "Second Thoughts on Paradigms." In *The Structure of Scientific Theories*, edited by Frederick Suppe, 459–82. Urbana: University of Illinois Press, 1974.

———. *The Structure of Scientific Revolutions*. Chicago: University of Chicago Press, 1962. 2nd ed. 1970. 3rd ed. with added index by Peter J. Riggs, 1996. 4th edition, with introduction by Ian Hacking, 2012.

————. Thomas S. Kuhn Papers. Institute Archives, Massachusetts Institute of Technology, Cambridge, MA.

Kuhn, T. S., and J. H. Van Vleck. "A Simplified Method of Computing the Cohesive Energies of Monovalent Metals." *Physical Review* 79 (July 1950): 382–88.

Kusukawa, Sachiko. *Picturing the Book of Nature: Image, Text, and Argument in Sixteenth-Century Human Anatomy and Medical Body*. Chicago: University of Chicago Press, 2012.

Lakatos, Imré, and Alan Musgrave, eds. *Criticism and the Growth of Knowledge*. Cambridge: Cambridge University Press, 1969.

Lederberg, Joshua. *"Escherichia coli."* In *Instruments of Science. An Historical Encyclopedia*, edited by Robert Bud and Deborah Jean Warner, 230–32. New York: Garland Publishing, 1998.

Lederman, Muriel, and Richard M. Burian, eds. "The Right Organism for the Job." *Journal of the History of Biology* 26 (1993): 233–367.

Lehoux, Daryn. *What Did the Romans Know?* Chicago: Chicago University Press, 2011.

Lewis, Clarence Irving. *Mind and the World Order*. Chicago: Scribners, 1929.

Lloyd, Geoffrey. *Ancient Worlds, Modern Reflections: Philosophical Perspectives on Greek and Chinese Science and Culture*. Oxford: Clarendon Press, 2004.

————. *Polarity and Analogy: Two Types of Argumentation in Early Greek Thought*. Cambridge: Cambridge University Press, 1966.

Löwy, Ilana. "The Strength of Loose Concepts—Boundary Concepts, Federative Experimental Strategies and Disciplinary Growth: the Case of Immunology." *History of Science* 30 (1992): 371–96.

Löwy, Ilana, and Jean-Paul Gaudillière. "Disciplining Cancer: Mice and the Practice of Genetic Purity." In Gaudillière and Löwy, *The Invisible Industrialist*, 209–49.

Lyons, J. D. *Exemplum: The Rhetoric of Example in Early Modern France and Italy*. Princeton, NJ: Princeton University Press, 1990.

Madsen, O. J., J. Servan, and S. A. Øyen. "'I Am a Philosopher of the Particular Case': An Interview with Holberg Prize Winner 2009 Ian Hacking." *History of the Human Sciences* 26 (2013): 32–51.

Markov, A. A. *Theory of Algorithms*. Translated by Jacques J. Schorr-Kon and PST Staff. Moscow: Academy of Sciences of the U.S.S.R., 1954.

Maslow, Abraham. *The Psychology of Science: A Reconnaissance*. New York: Harper and Row, 1966.

Masterman, Margaret. "The Nature of a Paradigm." In Lakatos and Musgrave, *Criticism and the Growth of Knowledge*, 59–89.

Mazauric, Simone. *Savoirs et philosophie à Paris dans la première moitié du XVIIe siècle: les conférences du Bureau d'adresse de Théophraste Renaudot, 1633–1642*. Paris: Publications de la Sorbonne, 1997.

Mersenne, Marin. *L'impiété des déists, athées, et libertins de ce temps*. Paris: Pierre Bilaine, 1624.

————. *Quaestiones celeberrimae in Genesim*. Paris: Sebastian Cramoisy, 1623.

————. *La vérité des sciences*. Paris: Toussainct du Bray, 1625.

Miller, George. "Psychology as a Means of Promoting Human Welfare." *American Psychologist* 24 (1969): 1063–75.

Mody, Cyrus, and David Kaiser. "Scientific Training and the Creation of Scientific Knowledge." In *Handbook of Science and Technology Studies*, rev. ed., 377–402. Cambridge, MA: MIT Press, 2007.

Montaigne. *The Essays of Montaigne*. Translated by E. J. Teichman. London: Oxford University Press, 1927.

Morange, Michel. "From the Regulatory Vision of Cancer to the Oncogene Paradigm, 1975–1985." *Journal of the History of Biology* 30 (1997): 1–29.

Mullins, Nicholas C. "The Development of a Scientific Specialty: The Phage Group and the Origins of Molecular Biology." *Minerva* 10 (1972): 51–82.

Naudé, Gabriel. *Advis pour dresser une bibliotheque.* Paris: François Targa, 1627.

———. *Apologie pour tous les grand hommes qui ont esté accusez de magie.* Paris: François Targa, 1625.

Nelson, Nicole C. "Modeling Mouse, Human, and Discipline: Epistemic Scaffolds in Animal Behavior Genetics." *Social Studies of Science* 43 (2012): 3–29.

Newman, William, and Lawrence Principe. "Alchemy vs. Chemistry: The Etymological Origins of a Historiographic Mistake." *Early Science and Medicine* 3 (1998): 32–65.

Nye, Mary Jo. *Michael Polanyi and His Generation.* Chicago: University of Chicago Press, 2011.

Ohme, Heinz. *Kanon ekklesiastikos. Die Bedeutung des altkirchlichen Kanonbegriffs.* Berlin: Walter De Gruyter, 1998.

Oppel, Herbert. *KANΩN. Zur Bedeutungsgeschichte des Wortes und seiner lateinischen Entsprechungen (Regula-Norma).* Leipzig: Dieterich'sche Verlagsbuchhandlung, 1937.

Park, Buhm Soon. "Disease Categories and Scientific Disciplines: Reorganizing the NIH Intramural Program, 1945–1960." In Hannaway, *Biomedicine in the Twentieth Century,* 27–58.

Piaget, Jean. *The Child's Conception of Movement and Speed.* Translated by G. E. T. Holloway and M. J. MacKenzie. New York: Basic Books, 1970.

———. *The Child's Conception of Physical Causality.* Translated by Marjorie Gagain. London: Routledge & Kegan Paul, 1930.

———. *Les notions de mouvement et de vitesse chez l'enfant.* Paris: Presses Universitaires de France, 1946.

Picardi, Mariassunta. *Le libertà del sapere: filosofia e "scienza universale" in Charles Sorel.* Naples: Liguori, 2007.

Pickering, Andrew. *The Mangle of Practice: Time, Agency, and Science.* Chicago: University of Chicago Press, 1995.

Pliny the Elder. *Natural History.* Translated by Harris Rackham. Loeb edition, 10 vols. Cambridge, MA: Harvard University Press, 1952.

Polanyi, Michael. *Personal Knowledge.* Chicago: University of Chicago Press, 1956.

Popper, Karl. "Normal Science and Its Dangers." In Lakatos and Musgrave, *Criticism and the Growth of Knowledge,* 51–58.

Rader, Karen A. *Making Mice: Standardizing Animals for American Biomedical Research, 1900–1955.* Princeton, NJ: Princeton University Press, 2004.

———. "Of Mice, Medicine, and Genetics: C. C. Little's Creation of the Inbred Laboratory Mouse, 1909–1918." *Studies in History and Philosophy of the Biological and Biomedical Sciences* 30C (1999): 319–43.

Reisch, George. "Anticommunism, the Unity of Science Movement and Kuhn's *Structure of Scientific Revolutions.*" *Social Epistemology* 17 (2003): 271–75.

———. "Did Kuhn Kill Logical Empiricism?" *Philosophy of Science* 58 (1991): 264–77.

———. "A History of the International Encyclopedia of Unified Science." Ph.D. diss., University of Chicago, 1995.

———. *How the Cold War Transformed Philosophy of Science.* Cambridge: Cambridge University Press, 2005.

———. "The Paranoid Style in American History of Science." *Theoria* 27, no. 75 (2012): 323–42.

———. "Planning Science: Otto Neurath and the International Encyclopedia of Unified Science." *British Journal for the History of Science* 27 (1994): 153–75.

———. "When *Structure* Met Sputnik: On the Cold-War Origins of *The Structure of Scientific Revolutions*." In *Science and Technology in the Global Cold War*, edited by Naomi Oreskes and John Krige. Cambridge, MA: MIT Press, 2014.

Renaudot, Théophraste. *Premiere centurie des questions traitées ez conferences du Bureau d'adresse, depuis le 22 jour d'Aoust 1633. Jusques au dernier Juillet 1634.* Paris: Bureau d'Adresse, 1635.

———. *Seconde centurie des questions traitées ez conferences du Bureau, depuis le 3 novembre 1634. Jusques à l'11 fevrier 1636.* Paris: Bureau d'Adresse, 1636.

Rheinberger, Hans-Jörg. *An Epistemology of the Concrete: Twentieth-Century Histories of Life.* Durham, NC: Duke University Press, 2010.

———. "Experiment, Difference, and Writing. Part 1. Tracing Protein Synthesis." *Studies in History and Philosophy of Science* 23 (1992): 305–31.

———. "Experiment, Difference, and Writing. Part 2. The Laboratory Production of Transfer RNA." *Studies in History and Philosophy of Science* 23 (1992): 389–422.

———. *Toward a History of Epistemic Things: Synthesizing Proteins in the Test Tube.* Stanford, CA: Stanford University Press, 1997.

Rigolet, François. "The Renaissance Crisis of Exemplarity." *Journal of the History of Ideas* 59 (1998): 557–63.

Ritzer, George. *Sociology: a Multiple Paradigm Science.* Boston: Allyn and Bacon, 1975.

Rodagem, François-Marie. "Sagesse Kirundi; proverbs, dictons, locutions usités au Burundi." In *Annales du Musée Royal du Congo Belge. Série in 8o.* Vol. 34 (1961). *Sciences de l'Homme.* Expanded in *Paroles de sagesse au Burundi.* Leuven: Peters, 1983.

Rouse, Joseph. "Science as Practice: Two Readings of Thomas Kuhn." In *Knowledge and Power: Toward a Political Philosophy of Science*, 26–40. Ithaca, NY: Cornell University Press, 1987.

Rudolph, John. *Scientists in the Classroom.* New York: Palgrave Macmillan, 2002.

Schaffer, Frederick L., and Carlton E. Schwerdt. "Crystallization of Purified MEF-1 Poliomyelitis Virus Particles." *Proceedings of the National Academy of Sciences* 41 (1955): 1020–23.

Scheffler, Robin. "From Polio to p53: The Life of Simian Virus 40." Paper presented at the History of Science Society Meeting, San Diego, CA, November 16, 2012.

———. "Managing the Future: The Special Virus Leukemia Program and the Acceleration of Biomedical Research." *Studies in History and Philosophy of Biological and Biomedical Sciences* 48 (2014): 231–49.

Schelling, Thomas C. *The Strategy of Conflict.* Cambridge, MA: Harvard University Press, 1960.

Schrecker, Ellen. *No Ivory Tower.* New York: Oxford, 1986.

Shapere, Dudley. "Review of *The Structure of Scientific Revolutions*." *The Philosophical Review* 70 (1964): 383–94.

Shapin, Steven. *The Scientific Revolution.* Chicago: University of Chicago Press, 1996.

Shapiro, Edward S., ed. *Letters of Sidney Hook.* Armonk, NY: M. E. Sharpe, 1995.

Sismondo, Sergio, et al. "Special Section: Fifty Years of *The Structure of Scientific Revolutions*, Twenty-five of *Science in Action*." *Social Studies of Science* 42, no. 3 (2012): 415–80.

Solomon, Howard M. *Public Welfare, Science, and Propaganda in Seventeenth Century France: The Innovations of Théophraste Renaudot.* Princeton, NJ: Princeton University Press, 1972.

Sorel, Charles. *De la perfection de l'homme*. Paris: Robert de Nain, 1655.

———. *La science universelle, tome quatriesme*. Paris: Theodore Girard, 1668.

Stanley, Wendell M. "Isolation of a Crystalline Protein Possessing the Properties of Tobacco-Mosaic Virus." *Science* 81 (1935): 644–45.

———. "Viruses and Cancer." Lecture presented at the Scientific Session of the American Cancer Society, New York City, October 22, 1962. Stanley Papers. Bancroft Library, University of California, Berkeley.

Star, Susan Leigh, and James R. Griesemer. "Institutional Ecology, 'Translations' and Boundary Objects: Amateurs and Professionals in Berkeley's Museum of Vertebrate Zoology, 1907–1939." *Social Studies of Science* 19 (1988): 387–420.

Stern, David. "Sociology of Science, Rule-following and Forms of Life." In *History and Philosophy of Science. New Trends and Perspectives*, edited by Michael Heidelberger and Friedrich Staedler, 347–67. Dordrecht: Kluwer, 2002.

Stevens, S. S. "Edwin Garrigues Boring." *Biographical Memoirs of the National Academy of Sciences* (1973): 41–76.

Stopes-Roe, H. V. "Review of *The Structure of Scientific Revolutions*." *British Journal for the Philosophy of Science* 15 (1964): 158–61.

Strasser, Bruno J., and Soraya de Chadarevian. "The Comparative and the Exemplary: Revisiting the Early History of Molecular Biology." *History of Science* 49 (2011): 317–36.

Stratton, George, M. "Some Preliminary Experiments on Vision without Inversion of the Retinal Image." *Psychological Review* 3 (1896): 611–17.

Temin, Howard M. "The DNA Provirus Hypothesis." *Science* 192 (1976): 1075–80.

Temin, Howard M., and Harry Rubin. "Characteristics of an Assay for Rous Sarcoma Virus and Rous Sarcoma Cells in Tissue Culture." *Virology* 6 (1958): 669–88.

Thorne, James P. "Review of Paul Postal's *Constituent Structure: A Study of Contemporary Models of Syntactic Description*." *Journal of Linguistics* 1 (1965): 73–76.

Toulmin, Stephen. *The Uses of Argument*. Cambridge: Cambridge University Press, 1958.

University of Chicago Press Records, box 278. Special Collections, University of Chicago Library.

Van Vleck, John H. John H. Van Vleck Papers. Department of Physics Collection. Harvard University Archives, Cambridge, MA.

Verdier, Gabrielle. *Charles Sorel*. Boston: Twayne, 1984.

Vidal, Fernando. *Piaget Before Piaget*. Cambridge, MA: Harvard University Press, 1994.

Watson, James D. *Molecular Biology of the Gene*. New York: W. A. Benjamin, 1965.

Weber, Max. *The Methodology of the Social Sciences*. Translated and edited by Edward Shils and Henry A. Finch. New York: Free Press, 1949.

Webster, John. *Academiarum examen, or the Examination of the Academies*. London: Giles Calvert, 1654.

Wellman, Kathleen A. *Making Science Social: the Conferences of Théophraste Renaudot, 1633–1642*. Norman: University of Oklahoma Press, 2003.

Werner, Heinz. *Comparative Psychology of Mental Development*. 2nd ed. with foreword by Gordon Allport. New York: Follett, 1948.

White, Ray, and C. Thomas Caskey. "The Human as an Experimental System in Molecular Genetics." *Science* 240 (1988): 1483–88.

Witkin, Herman. "Heinz Werner, 1890–1964." *Child Development* 36 (1965): 306–28.

Wisdom, John. "The Logic of God." In *Paradox and Discovery*. Oxford: Blackwell, 1965.

Wittgenstein, Ludwig. *Philosophical Investigations*. Translated by G. E. M. Anscombe. 3rd ed. Englewood Cliffs, NJ: Prentice Hall, 1958. 3rd rev. ed. Oxford: Blackwell, 2001.

Wray, K. Brad. "Kuhn and the Discovery of Paradigms." *Philosophy of the Social Sciences* 41 (2011): 380–97.

Wright, Susan. *Molecular Politics: Developing American and British Regulatory Policy for Genetic Engineering, 1972–1982*. Chicago: University of Chicago Press, 1994.

Yerxa, Donald A. "Introduction: Historical Coherence, Complexity, and the Scientific Revolution." *European Review* 15 (2007): 439–44.

Yi, Doogab. "The Enemy Within? Oncogenes and the Demise of the Special Virus Cancer Program in the 1970s." Paper presented at "Debating Causation: Risk, Biology, Self, and Environment in Cancer Epistemology, 1950–2000," Princeton University, October 21, 2011.